Published 23 October 2009

New version 1 August 2010

Fractal on cover
Hypercube Logo
and other graphics
outputs from
QuickBASIC programs

What Happened . . .

IT IS VERY PUZZLING that the most important problem in the world is without attention, hardly talked about, just to mention it would be "not politically correct".

It was without mention in the recent Copenhagen Conference.

Is this a symptom of the essential human final defeat?

Is not the general commotion about climate change rather a distraction?

This book gives a new opening for the subject.

By S. N. Afriat

Production Duality and the von Neumann Theory of Growth and Interest. Meisenheim am Glan: Verlag Anton Hain, 1974. Pp 86. *Mathematical Systems in Economics*, 11.

STUDIES IN CORRELATION: *Multivariate Analysis and Econometrics* (with Gerhard Tintner and M. V. Rama Sastry). Göttingen: Vandenhoeck and Ruprecht, 1975. Pp 150. (Contributing Part I: The Algebra and Geometry of Statistical Correlation. Pp 100). *Angewandte Statistik und Ökonometrie.*

Combinatorial Theory of Demand. London: Input-Output Publishing Co., 1976. *Occasional Paper* No. 1.

THE PRICE INDEX. Cambridge University Press, 1977. Pp xvi + 187

Demand Functions and the Slutsky Matrix. Princeton University Press, 1980. Pp xii + 269. *Princeton Studies in Mathematical Economics*, 7.

The Ring of Linked Rings. London: Gerald Duckworth & Co Ltd, 1982. Pp xviii + 126. (Mathematical and computer recreations, theory and history of Chinese Rings, and its various manifestations, binary dividers, Lucas's Tower problem, error correcting code, Dragon Curves, &c.)

Logic of Choice and Economic Theory. Oxford: Clarendon Press, 1987. Pp xiv + 577.

LINEAR DEPENDENCE : *Theory and Computation.* Kluwer Academic / Plenum Publishers, 2000. Pp xiii + 169.

The Market : equilibrium, stability, mythology. Foreword by Michael Allingham. Routledge, 2002. Pp xv +128.

The Price Index and its Extension—A chapter in economic measurement. Foreword by Angus Deaton. Routledge, 2004. Pp xxx + 421

on the River Non. Foreword by Paolo Vivante. Empoli (FI): Ibiskos di A. Ulivieri, 2006. Pp. xvi + 33. (Twenty-four poems, one watercolour, graphic outputs from QuickBASIC programs, and other items.)

On the latent vectors and characteristic values of products of pairs of symmetric idempotents. *Quart.J. Math. Oxford* 2, 7 (1956), 76-8.

The Cube. In *Art Has Many Facets*, edited by J. MacAgy, Fine Arts Department, University of St. Thomas, Houston, Texas, 1963.

People and Population. *World Politics* 17, 3 (1965), 431-9. Japanese translation, with foreword by Edwin O. Reischauer (US Ambassador to Japan): *Japan-America Forum* 11, 10 (October 1965), 1-28.

Graphic A-Mazes. *Creative Computing* 6, 6 (June 1980), 124-7.

On the constructibility of consistent price indices between several periods simultaneously. In *Essays in Theory and Measurement of Demand: in honour of Sir Richard Stone*, edited by Angus Deaton. Cambridge University Press, 1981, 133-61.

His work is a classic illustration of how much we learn from new ways of thinking.

Angus Deaton

Overjoyed with ... documents testifying to your universal genius: poet, painter, economist, mathematician, biographer.

Paul Streeten

Sydney Afriat is famous for his unique and penetrating insights, often very unsettling to those who have worked long and hard in a field—without ever seeing what is obvious to Sydney, who then formulates it neatly in very compact mathematics.... will very likely render many current discussions ... obsolete.

Edward Nell

Sydney Afriat belongs to that select group of economic theorists who have become a legend in their own times. There is such a thing as *"Afriat's Theorem"*, which has become part of the staple for students of microeconomic theory ... Moreover, he may belong to another select group of prose stylists who are also masters of some aspects of the mathematical method and its philosophy.

Vela Velupillai

Dedicated

to

James Lovelock

Pericles favoured "the intelligence which proceeds not by hoping for the best (a method only valuable in desperate situations), but by estimating what the facts are, and thus obtaining a clearer vision of what to expect"

James Lovelock has lead in "estimating what the facts are"

What Happened
to
The Population Problem?

—and other questions

S. N. Afriat

Bright Pen
www.authorsonline.co.uk

ISBN: 978-0-7552-1190-6

s.afriat@gmail.com

A Bright Pen Book

Authors OnLine Ltd
19 The Cinques
Gamlingay, Sandy
Bedfordshire SG19 3NU
England

This book is also available in e-book format,
details of which can be found at

www.authorsonline.co.uk

Contents

Preface

IT IS VERY PUZZLING that the most important problem in the world is without attention, hardly talked about, just to mention it would be "not politically correct".

It was without mention in the recent Copenhagen Conference.

Is this a symptom of the essential human final defeat?

Is not the general commotion about climate change rather a distraction?

This book gives a new opening for the subject.

ONCE THERE WAS something called "The Population Problem". The term had been household language for a well recognized matter. After attention to the subject, decades passed. Then there came a point when it appeared the 'Problem' had vanished! Instead of abundance of speeches and writings, references to it could hardly be found. There was some mystery in the silence. What had happened to the 'Problem'? was a good question.

At an event at the university in Rome, *Laurea Honoris Causa* for Angus Deaton, June 2007, there was some discussion about 'What Happened?'. The question brought the comment from Deaton "It's coming back" and there was recollection of R. D. C. Black who had been Editor of *World Politics* in Princeton in 1965 when my essay "People and Population" had been published. This book amounts to response to his letter that came later. These exchanges took us back to those early days and stirred up the 'What Happened?' question.

Mention at a meeting of the Cambridge Society, Tuscany & Umbria, brought guidance from Andrew Davies that now any reference to 'the Problem' would be "Not Politically Correct". He sent me the Nicolson-Lord article in the *Guardian Weekly*, then from Andrew MacMillan I have various documents like "Return of the Population Growth Factor" and attention to the Partha Dasgupta and Boris Johnson articles. I thank these for guidance for opening an approach to "What Happened?". Touched on also is the interesting new 'World Family' WF concept that may have influence for population, and for poverty and perhaps other factors.

Fallout from climate change and economic crisis have shaken the system and are bound to affect ideas. It therefore seemed suitable to include the two sections about unsettled parts of economic teaching, one concerned with the 'welfare' theory of the market and its absurdity, the other with ideas of early writing of J. M. Keynes that have become displaced by opposites whose time seems now to have come.

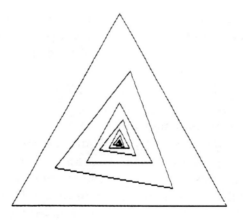

Acknowledgements

For materials reproduced, thanks are due to the Princeton University Press for permission to reproduce the following:

People and Population. *World Politics* XVII, 3, April 1965, 431-9

My thanks for valued exchanges are due to:

Michael Allingham, Mary Ann Beckinsale, Angela & Korhan Berzeg, Mario Bunge, Patrick Creagh, Julie Daniels, Andrew Davies, Angus Deaton, Ali Doğramacı, Richard Falk, Luca Fiorito, Victor Funnell, Edward Goldsmith, Katharina Hietarinta-Zonnekein, Janne Hoff-Tilley, Nuri Jazairi, Laya Labi, Axel Leijonhufvud, Christopher Ligota, Fausto Lore, James Lovelock, Luigi Luini, Andrew MacMillan, Jacinta Nadal, Edward Nell, Guinevere Nell, Pringle sister *et al*, John Alan Robinson, Nate Rosenberg, Anna Scaife, Lidia Sciama, Susan Senior, Martin Shubik, Norman Stone, Paul Streeten, Jenny Stringer, Vela Velupillai, Cesare Vivante, George Woodcock, Sean Wylie, Giulio Zanella

I should add Alan and Gwen Robinson for leading me to Authors OnLine, and Richard Fitt for distinct satisfaction of being an OnLine Author

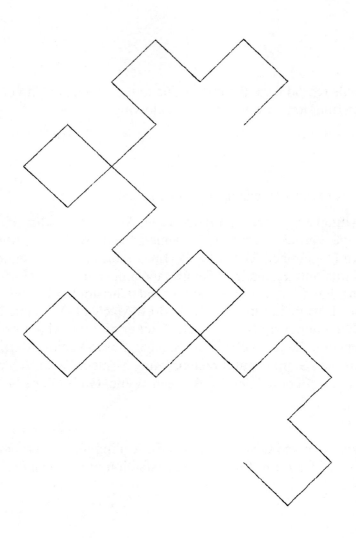

What Happened . . .

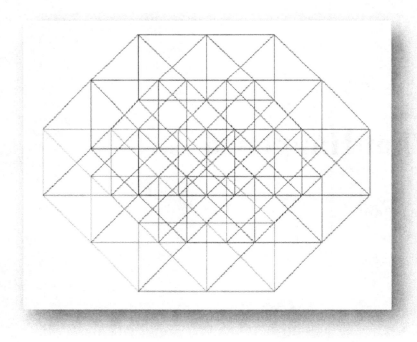

1
Vegetarian Pie in the Sky

About the recent UN vegetarian directive
and a remark of Einstein half-century ago

> **"Nothing will benefit human health and increase chances for survival of life on earth as much as the evolution to a vegetarian diet."**
> **Albert Einstein**
> Quote of the Day

This remark issued as "Einstein Quote of the Day"[1] by a Google computer gadget must have been made more than a half-century ago.

It is hard to live with this device which is best disabled after a short run. Einstein is a great authority in physics but his everyday remarks are not so stimulating. However, Einstein is not noted for frivolity; the remark compels a pause. From another source it might be put aside without further thought whereas now it has to be puzzled about[2].

What could Einstein have had in mind? Prejudice against cannibalism extended to feeding on our more distant cousins? Some might like that idea. And what time horizon? "Evolution" suggests centuries, or millennia.

---

[1] Phillip Olsen, *Albert Einstein Quote of the Day*, phillip132@gmail.com.
[2] In Princeton at the house of Eric Kahler, friend of Einstein who died the year before, whose step daughter and secretary would visit, there was some mention of his vegetarian lifestyle.

After having put the remark aside for a long time, I was amazed to view in a newpaper shop:

"UN says eat less meat to curb global warming. Top expert urges radical shift in diet."

The Observer
Cover page headline
Sunday 7 September 2008[3]

Though no regular newspaper reader I did get this, and on arriving home promptly reinstalled "Einstein Quote of the Day" hoping to get light on his thought from another Quote.

The coincidence is a surprise and it seemed as if here exposed was a less familiar dimension of Einstein's well known genius. He had thrown out the remark a half century ago without any of the urgency and elaborate foundational researches of the UN experts. What was his foundation?

After some days came the following Quote:

"It is my view that the vegetarian manner of living by its purely physical effect on the human temperament would most beneficially influence the lot of mankind."

The approach of Einstein seems therefore broadly different from that of the UN experts. All the same it makes no conflict and rather adds support.

While the UN group went on as usual with grief about Global Warming and Saving The Planet, his own focus was more on Saving Mankind, as it were the software instead of the hardware. The UN approach has earlier history in Australia and New Zealand where the hole in the Ozone Layer was connected with the cattle and sheep presence there[4].

Then a later Quote:

---

[3] Robin McKie and Caroline Davies, Special Report, *The Observer*
[4] Guided by Australian friend Jennifer Storey.

"Few people are capable of expressing with equanimity opinions which differ from the prejudices of their social environment. Most people are even incapable of forming such opinions."

It should be useful to have that available for reference, for which purpose it is named *Einstein's Theorem*.

Top importance is given nowadays to Global Warming, with recurring pronouncements about the Carbon Footprint of anything. A half century ago there was The Population Problem having attention[5], about which *political correctness* now requires *silence*[6]. While Globalism is the current orthodoxy mentioned mindlessly by every respectable politician, that most respected and brilliant economic leader J. M. Keynes used to preach the opposite[7], maybe for then, and for now, completely right. At any time the unquestioned endlessly mentioned prize is always *growth* without hint there could come a time when 'growth' is a *bad* word. Respect for the issue of poverty has been overwhelmingly unsuccessful and masses of the destitute have additionally multiplied from disasters, natural, economic, and war. They may have to take over the world, as did the famous marauding hordes and sackers of cities of another age—after all, what else? Einstein's Theorem amplifies dismay that may be felt about the future.

The UN Experts have brought into the open a perplexed complex of issues to join what we already have. The many entanglements prevent a simple overview not now required after the report of Robin McKie and Caroline Davies in *The Observer*.

---

[5] For "People and Population", *World Politics* 1965, the translation into Japanese and commentary by US Ambassador may, it seems, have had some effect, for the Japanese government now concerned about low birthrate has shortened working hours to promote family life. A highly unsuitable action considering that to live in a place where population is falling should be a highly appreciated privilege.

[6] I am indebted to Andrew Davies for pointing this out to me after observation of the long time since mention of the Problem and then providing the 2007 article by David Nicholson-Lord in the *Guardian Weekly*, and to Andrew MacMillan for other documents and guidance.

[7] As reported here in Part 5 "On Trade, and Self-sufficiency".

There is relief to be told governments should not "regulate", and alarm that farmers in 2006 produced four times as much meat as they did in 1961. From UNFAO "nearly a fifth of all greenhouse-gas emissions come from livestock, more than from all forms of transport"[8]. The modest "urging of the world to give up meat for at least one day a week" is, it may be submitted, no great violence to "those who believe the lifestyle of a carnivore is perfectly acceptable, both morally and environmentally". Some of us may have gladly gone along with such an idea already. Weekend days being already loaded, why not Wednesdays? Now there can be wonder what prospect there is for progress with this manifestly important issue, not "incorrect" to mention, or whether it may become the Vegetarian Pie in the Sky.

---

[8] From James Lovelock " ... about 24% of human CO2 emissions come from the metabolism of people, their pets and livestock? Removing the pets and livestock would reduce the emission to 4%."

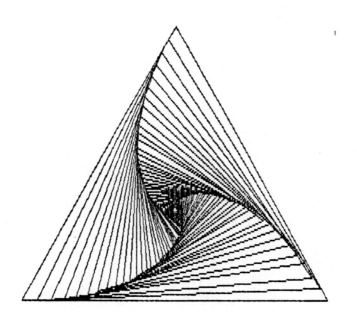

2

What Happened to
The Population Problem

The emergence of the 'What Happened?' question is set forth in the Preface.

Affecting this project, to be taken up now, there have been surprising experiences on TV, beside more personal ones.

Among programmes that record fire, flood and disaster there was one about poverty. A key moment was showing a man who did not have enough to feed his twelve children. There was compassion but no comment on wisdom of having so many children you cannot feed. Anyway, it appeared now 'Not Socially Incorrect' to have any number of children.

A similar case was a programme about social good works surely to inspire others to follow. There was report about a stately good works lady, a pillar of society model, and one could only hear with wonder and admiration the list of her good works. At the end there was mention she had twelve children (twelve also in last case just a coincidence). Maybe she could feed them, but again there was no hint at any idea of the incorrectness of having so many children; rather the contrary if anything here was a crowning achievement being marked. One might well be surprised, and shocked.

That is enough about TV, now instead more about personal friends perhaps now not quite as friendly as they had been.

I visited an old friend whose son I believed to have two children. When I asked Mother about Son I learnt he now had four.

Unfortunately, beside surprise I must have let slip some indication that I took that to be many. Mother seemed offended and explained, as it were in defense of Son, that population was falling and more children were needed to look after us when we got old. I found that an unpraisworthy reason but said nothing. I now gather an argument like that had been circulating.[9] My thought had been that influence more likely came through Son's wife, or—from nowhere ... after all, heightened awareness of "The Problem" had become depressed. A coolness crept into our relationship, even though a call from Mother when I again did not show up for some time made me feel perhaps we had got over it.

Now for a similar case, again with offence I hope not to be compounded by this report, already submitted for discussion with the individual involved, a pioneering ecologist from half-century ago who had been talking for years the way the most vocal do now (are these also the ones with the "politically correct" silence about "The Problem"?) Anyway, he stopped by one day and put a large volume on my desk which I opened to see the dedication "To my children 1, 2, ..." It is an enterprising work not to be diverted by a dreary matter, but an exhibition of amusement could not be avoided. Then followed a long gap ...

That now is enough about friends maybe now not quite as friendly as they once were. They tell something simple like that absentmindedness, or just ignorance, might be a factor. But undoubtedly there is more to "What Happened" than just that.

Here there should be note that a large part of the history of doing anything about population has been so dreadful as to silence everyone.[10]

---

[9] When extra children to look after the old have grown up they will need extra children to look after them when they get old and so forth, under ordinary conditions producing compounded exponential population growth so that after some generations there would be no room.

[10] See for instance Matthew Conolly, *Fatal Misconception: The Struggle to Control World Population,* Harvard 2008, reviewed by Dominic Lawson, *The Sunday Times,* 18 May 2008. But still, the distinguished population policy in China, a drastic government imposition, has been broadly well understood and accepted as necessary, and a benefit to the people, and to the World.

After rejection with just about every recognized approach, and without such an approach for reference, there is no real point to making any mention of the matter at all. Hence the silence may not entirely be a matter of "political correctness" but of common sense. As it were, there is no point in talking about the matter because there is nothing to do about it. That is, even though the simplest intelligence puts the population problem as riding high and as a most central point with everything of importance, to the extent that without some success with it any other achievement must be futile.[11]

But of course, there may also be complex politics to the matter, not now to be unraveled..

That appears to be the pitiful situation today.

Half-century ago when the problem had attention, and since, there was the unspoken assumption that it would have reference to a region and anything to be done about it would be responsibility of the government.

The record of government initiatives fostered the view the problem was hopeless and nothing could be done about it. The consequent scarcity of references caused the decline of awareness and created the impression there was no such problem. This is confirmed by cases of ignorance encountered everywhere, as just now illustrated.

But the real importance of the problem has had no decline with its neglect; on the contrary, with emergence of the 'green' movement and concern about climate change, it has taken on a newly explicit massive importance.

This is a broad view of "What Happened . . . " Let us now consider whether the problem is hopeless or not, the approach should be local or global, and so forth.

---

[11] Population Institute (1990). No matter what your cause—*it's a lost cause*—if we don't come to grips with overpopulation. *The Population Institute*, Washington DC. Boris Johnson (2007). Global over-population is the real issue. *Boris Johnson's Office* (he is Mayor of London). Yes, of course, but what can or should be done about it?

Though there are alternatives, here adopted is the simple view that the approach should be global without reference to restricted regions, also involve all people of the world equally together

A familiar observation is that families tend to be smaller in advanced countries, so populations can even be falling. From this there can be the conclusion that the "Population Problem" is not an inalienable human characteristic, and not essentially and fundamentally hopeless.

Also from this it may be argued that all that needs to be done is simply to assist countries to develop. Unfortunately, development has no guarantee, in any case takes time, and can be negative.

There cannot on this score be relaxation of the need for high worldwide awareness on the part of all persons. The current deficiency has to be remedied by provision of education giving knowledge of the realities, and of instruction and means for response. The approach to the problem should be broadcast worldwide and with that joined with information and means for Family Planning provided freely everywhere, with appropriate administration.

That, for a most elementary practical approach to population, there never has been such a provision, does suggest pronounced pronatalist resistances in the matter. Wanting drivers for their tanks, Mussolini and Hitler were pronatalist. They are gone now, but still there are others.

Lacking for the broadcast are easily reproduced well prepared statements translated into all languages. For a start immediately let us say a single page, on one side of a sheet and hence easily copied, then later two sides, and so forth subject any time to improvement.

Called for just now outstandingly is the simple single page. Then looming large for anyone pretentious enough to write all this is, perhaps, the obligation to offer a model for the statement. Or better, get some enthusiastic other person to do that. A prize should be offered for the best submission.

"Political Correctness" may usually have lesser importance than the all conquering "Social Correctness". Hence let it be recognized that, while today it may be determined as "Not Politically Correct" to mention the "Population Problem", tomorrow it may well be judged "Not Socially Correct" to show disregard for it.

I followed a BBC Hardtalk with Gerd Leipold of Greenpeace International. From brief mention, population increase appeared taken for granted without hint of influence to disturb it. The control of climate change seemed the generally accepted thing to have attention. Here was a classic instance of "politically correct" silence about population.

It was without mention that population increase could put beyond us the needed climate control, that may be to some extent already beyond us anyway.

But what about population? That cannot be beyond us, it is us! So in principle, population increase could be stopped instantly, with the right attutude. Surely we have to wake up about that again like half-century ago. Signals of every kind coming from everywhere insist on that.[12]

So never mind about Greenpeace. How about Lovelock, and Gaia?

> "It is not the carbon footprint alone that harms the Earth; the peoples footprint is larger and much more deadly."

James Lovelock
The Vanishing Face of Gaia: A Final Warning (2009, p. 75)

In fact the 'Population Problem' has been with us and well recognized quite apart from and even long before any thought of climate change and carbon:

> There was a time when numberless races of men wandered the earth. Seeing this Zeus took pity and resolved in the wisdom of his heart to relieve the all-nourishing Earth of men, stirring up the great quarrel of the Trojan war in order to lighten the burden by death. The heroes perished in Troy and Zeus' plan succeeded.

The Cypria, attributed to **Stasinos**, *ca.* 7-5th century BC[13]

---

[12] For example the Special Report "The End of Plenty" by Joel K. Bourne Jr, *National Geographic Magazine*, June 2009, http://ngm.nationalgeographic.com/2009/06/cheap-food/bourne-text.
[13] I am indebited to Christopher Ligota, Warburg Institute, for this quotation.

So now the old 'Population Problem' has importance intensified by a new climate dimension. But instead the effect of the new dimension is to have the old problem' put aside and attention given instead just to its new dimension, serving then as a help to forget about the old problem. Absurd! But the matter of climate change is penetrated that way; mountains of data to drown everyone, mountains of discussion to confuse everyone. It is better to have clear and simple Population as fundamental and the first thing for attention,

In the earlier condition of life there was a Population Problem. Single-cell plants, photo-synthesisers, "so grew and multiplied two billion or so years ago" as to change their world. A range of ecosystems were "condemned to an underground existence".

> "Their pollution was oxygen, a poisonous, carcinogenic, and fire-raising gas that life, including us, has evolved to benefit from."

Really! This new chapter for *The Origin of Species* must make Lovelock to Darwin as Einstein is to Newton,

The spleandour of Lovelock's view is filled out in his Gaia Theory, where Earth's atmosphere appears biologically produced and regulated, serving him as an original leader in the complex matter of emissions to the atmosphere that induce climate change.

Here, with concentration rather on the population problem, we cannot treat further that subject that has unmatched account from its originator

To follow now is some account of the 'World Family' WF concept serving for play with issues to do with Population, Poverty, and Peace. It is submitted for examination for its potential, if any, as a tool.

There is need now to explore how an approach to population could meet wanted objectives and avoid objectionable features. Some such may go loosely like the following, that may have appearance of representing an approach, or perhaps the start of one.

For a simple, essential and easily understood beginning, let us simply say people have the right to self-replacement, to replace themselves but not to exceed that. In other words, usually, a couple may produce two children, and no more. The "usually" here is inserted to escape rigidity and open the way for important departures.

This Replacement Principle RP, precise and simple, is easily understood by everyone. It serves better than any more complex, or nebulous, entertainment. It could be submitted as central for a movement; or needed new humanity. That is more or less the argument for the 1965 essay reproduced in the next section, at that time maybe fanciful without power for bond with reality.

However, alerted as we are now to the alarming future prospect, awakened as we have become of our part in destruction of our world, early by Lovelock and lately in his new book[14], times have changed.

But sleep is not to be disturbed, business with calculations with the popular Carbon Calculus measuring footprints everywhere, helps protect it. There really is just one factor that asks for treatment before any other, the last to get any mention, if any at all . . .

Lovelock somewhere compares the situation with war when, as some may remember, one really should ask if some project, possibly just a trip in the car, is really necessary. Then it could perhaps be seen as a World War in which we all happen to be on the same side, and to be the Enemy, and Defeat to be Victory.

Beside RP let it be submitted that this movement, though admitting assistance or support from governments, should essentially not be regarded as to do with government regulation. Goverments are all sorts, and ideally none should have dominance. In any case, a government has command in a restricted region, whereas the regard here is global, without reference to regions and without special allowance for any particular region, or people.

So far we have the idea that the approach to population should have simplicity, and, but for providing some service, fundamentally not be the business of government, hence voluntary, with extra dimensions that provide appeal.

---

[14] *The Vanishing Face of Gaia: A Final Warning.* New York: Basic Books, 2009.

Envisage a collection of people all voluntarily committed to the replacement principle RP, and with that at the same time are dedicated to mutual support in some form, as it were somewhat like members of a 'family', the 'World Family' WF. Individuals may join if the idea appeals to them, or not. Rather than stand alone before the promised disasters, better to have company.

Let us here entertain an idea that broadens the foregoing with an additional principle, expanding the "usually" qualification that was inserted along with RP. If a couple wish to be parents to an additional child beyond the usual two, another couple should give up in their favour one child from their two, by voluntary agreement. But there can be couples who give up having children anyway so that, but for recruits, WF numbers would fall.

For a basis for such agreement there may be various possibilities. A simple case is where the couple with one more child compensates the one with less, with some material, possibly just financial, support. The one couple quite likely is rich and the other poor. This feature therefore incorporates a treatment of 'poverty', that for ever has comment and is for ever elusive of actual treatment. Ultimately there should be no individual in WF without secure food and shelter, as may be with some needed refreshment of economics. In old civilizations land was the real wealth and everyone had some, a good plan deserving some new consideration.

The particular value and significance of WF is that it should be the carrier of fragile civilization, through the now promised times of destruction when otherwise it would surely have no survival. But such times are already not so distant, as exposed by any session of TV Daily News.

Now we are told that of six-billion populating the Earth only half-billion may remain at end of the century. So why worry? If anything more children are needed! What a relief!

But what needs to survive is not population but civilization. This is the point about WF. It should be like a boat that will float through the havoc of climate change, economic, or whatever.

Now there is the spectacle of large families living on handouts from UN and other agencies. Some procedure such as registration with WF could be required. And so forth with a variety of similar powerful roles for WF, making membership an important consideration for many.

Observed so far is the role of WF, a new concept, potential unknown, for treatment of population, and of poverty. Now to be remarked is the potential of WF as a factor for peace.

This should depend on members being numerous, scattered freely throughout the world and opposed to violence, so eventually evolving into a significant general opposition.

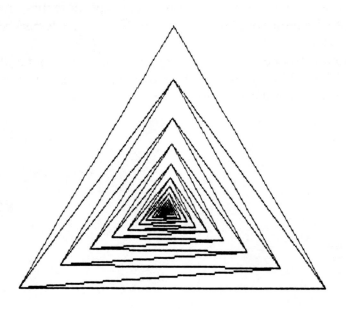

3
People and Population

World Politics **XVII, 3, April 1965, 431-9**

Japanese translation with foreword by Edwin O. Reischauer
(US Ambassador to Japan):
Japan-America Forum **11, 10 (October 1965), 1-28**

Abstract: *International Political Science Abstracts UNESCO*

T HE human population is the base of human existence. This, when considered, seems an inescapable proposition. But in the minds of most men it must have an absurd remoteness. The process of population has seemed like primordial creation itself, hardly a thing to be touched by mere scruple. Beyond the scope of any deliberation, the peopling of the earth has, for mankind as a whole, remained unquestioned as the earth, an axiom behind which there was no going and from which all proceeded. But according to a relentless gathering of awareness, as witnessed in an abundance of recent writings, it appears that history has marched to a new point. The condition of the world is being modified by a constellation of emergencies, and any observation and reflection on them, any rough glance at the outlines of trends in the life of the world, points to the swelling flood of population as the central reality in every perspective on the future.

The astonishing statistics make a now familiar opening alarm in numerous reports on the "population explosion" and need no further repetition. What has been pointed out especially is the appearance of another formidable curse in the world, of the order of the thermonuclear bomb. But another view is that here is a prompting, come in good time. As a parallel, the prize for successfucl confrontation of the existence of the bomb is ultimate military peace, a prize which would, so to speak, be the gift of the bomb—one which experience seems to suggest could not be given by any other authority. So also it can be proposed that the looming population crisis is a spur to greatly desired ends which mankind has not the fiber to approach otherwise. The confrontation of the population crisis provides the handle for taking hold of a complex of fundamental problems: it provides the compulsion and the cardinal point for approaching these problems as a whole. Without such a confrontation, it can be contended that there is no secure base for approaching these problems at all, because their entire ground is out of equilibrium in a fundamental respect.

It is in any case clear that, great as has been traditional complacency in the matter of population, this newly imposed uneasy questioning will now be undertaken with reluctance. It will bring into dispute the always accepted liberty of parenthood, where from the beginning there has existed an accord founded on general unconcern. Certainly there are codes which have had bearing on population, as we all know. But they look another way, and their interest has been with principles relating to circumstances of the act of reproduction, and not especially to the outcome, and, if anything, they have defended the growth of population. The tables may now be turned in a world as different from the old as is the geographer's globe from the flat which stretched apparently without limit round the ancients.

Among the elements of change in the world, any one of which is sufficient in itself to bring about a revolutionary transformation, certain things are outstanding: the attainment of the effective limit of destructiveness in the weapons of war; the automation of labor, with the possibility of massive replacement of individual human functions by efficient mechanical substitutes, with far-reaching consequences in every phase of life, along with the limitless advance of communications and data technology, and the advent of the computer; and the "population explosion". The response which is called for challenges the entire human race. In terms of problems and solutions, not only are answers hard to find; it is more particularly hard to know what are the proper questions. The scientific and statistical studies to which the enquirer easily makes recourse, and hopes to find nice objectivities, do not suffice. The realm of values is exposed—the realm we like best to leave alone, where practical reason itself has an essential point of departure. But it is certain that, as there is a general human will to survive and to pursue the values of life, these questions in all their difficult intricacies and profundities will have to be explored, with the curiosity and pain inevitable in so vast and uneasy an enquiry.

As with an infant, so with the progress of the race, at first there are only immediacies, and no questions. But as growth leads into the world, touching further boundaries and making for new relativities, existing equilibrium is continually lost in emergencies, and it is continually regained only through new knowledge and adaptation, in laws and devices which are increasingly indirect in their cause. In the moral sphere, the focus is usually not so much on the rational ground of a necessity, as on a principle which is traditionally taught, and has authority for the moulding of habitual behaviour, and may be reinforced by the contingencies and incentives which come from experience. However such principles originate and are maintained, they are a germ for the pattern of life, and they have inertia through generations. So far as procreation is concerned, as with morals for the general form and conduct of life, humanity is hardly out of primordial mother's arms. But the moment has come to go to school, or there can be no welcome for a new age. Primers for study include several recent works that touch aspects of the population matter.[15] They signal the kind of intensive attention that the question requires, the enlargement of thought that has to be persevered with and brought uppermost in the world's intelligence. As the title of one indicates, there is a dilemma. But the nature of the dilemma is not altogether explained. A certain dilemma does, however, appear when the gathering of evidence and opinion on negative effects of the "explosion," in three of the volumes, is brought together with the view, set out in the fourth, that massive population is the decisive source of great national power. This now is a doubtful view. But almost every facet of the subject is penetrated by controversy. It is impossible

---

[15] Philip M. Hauser, ed., *The Population Dilemma* (Englewood Cliffs, NJ 1963); Stuart Mudd, ed., *The Population Crisis and the Use of World Resources* (Bloomington, Ind. 1964); Katherine Organski and A. F. K. Organski, *Population and World Power* (NY 1961); and Melvin G. Shimm, ed., *Population Control: The Imminent World Crisis* (Dobbs Ferry, NY 1961).

to encompass in a simple statement the material in these volumes, with their many expert contributors, each dealing diversely with various facets of the subject, and making together something of an explosion of contribution to the great complexity of the matter. There is an impressive number of fingers in the population pie; economic, political, religious, military, medical, eugenic, cultural, educational, and so forth, with no clear limit. Whatever confrontation is going to be made to the problem will give an underpinning to the whole future existence of the human race, and the labor of this confrontation must certainty be one of the greatest intellectual and imaginative undertakings. The goal is convergence on a single resolve which is inherent in the situation. If a true dilemma does exist, if no such resolution can be reached, then the future is inevitably a hopeless surrender to overmastering forces which, it seems, must carry away things valued and pursued throughout ages.

There can be no submission to such a dissolution as a fatal inevitability. A focus on the process of population in its details and intricacies, such as is presented in the volumes which have been mentioned, is essential to enliven the consciousness. But when consciousness is roused, the approach cannot be made so much by expert factual and statistical exposition as in a realm of primitive essences. For the complexity is part of an intellectual entrenchment. But with the achievement of awareness it can fall away and leave a quite childish picture of the matter. Such is all that is wanted, and all that is fitting, and only a simple formula could make its way. Perhaps in another age, given the configuration of circumstances, an old man, in a cloud and on a mountain, surreptitiously chiselling stone, might have pulled it off. For the kind of restraint which must be entertained is not to be imposed, and perhaps capriciously played with by an arbitrary external authority, but a principle in process and in the historic moment of emerging with an innate validity.

A principle will be lost if it is not self-imposed and self-regenerating, by its own necessity. Such is the case with restraint on theft and murder. It is not dictated by a central authority; but it is in the fabric of every existing society.While it represents a tyranny over free impulse, at the same time it makes for liberation from a compulsive, absorbing involvement in threats from such impulses at large. It is a privation which is supported by virtue of a rewarding advantage, turning effort away from self-consuming exercises, such as preoccupied the proverbial caveman with his celebrated club, towards advantageous modes of endeavour. Civilization has advanced that way, by the suppression and at the same time by the release—in other words, the containment—of fundamental modes of aggression. Aggression is inseparable from life: not much could happen without it. So there can be no complete suppression of it, only its transformation, by restraint and compensating release. The suppression of aggression by theft and murder has been necessary to advance society, even an advancement demanded by competitive need; those societies have undergone subjugation or dissolution which did not achieve such supression. What could be effectively emerging into recognition is that the same is true of aggression that takes place by a reproductive process which is multiplicative—that is, which exceeds replacement. Once this excess was not self-destructive but was a means of confronting destructive threats from nature and man. But the raw impact of nature has receded greatly, and man, with his works and immediate presence, crowds around. He himself bears in self-attrition the brunt of this excess.

This is an inward edge to the matter, but there is another outer edge, where the reference is world-wide, giving the matter a comprehensive scope. For the question of population inwardly touches a society's direct self-interest and outwardly it touches forces of turbulence at large in the world. There is the completely explosive condition of the world, based on the unrest of miserable populations, vulnerable to the aggression which takes place by subversion, which has misery as the lever, and all this joined with the peculiar nature of existing armaments—concerning which no comment is needed, given the vivid and familiar picture furnished by sophisticated experts of thermonuclear war. Men always ask for material welfare and secure peace, and make continual reference to freedom. Disarmament is proposed merely because armaments are now at the extreme of destructiveness. But there must be a price: so what is it, and where is the readiness to pay it? It may turn out to be a bargain, as apparently was felt to be the case with theft and murder. Certain modes of conduct, relished in one age, become completely foreign in another. These days cannibalism is almost entirely eradicated from our impulses: our detachmentis so fine that we even look upon it with amusement, though at one time its surrender may have seemed to exact an unpleasant price. By the same token, in future days restraint on parenthood may become entirely accepted, even though today it may look like a taxation on natural rights against which many will rebel. Humanity is plastic and, given the temperature, takes the shape of the vessel into which it flows, as is shown in every facet of civilization. But a vessel need not be a prison; what is wanted is a confinement which is also a vehicle of escape.

To the need for welfare and disarmament, there seems no reply in the present order. We have found that weifare gained in one dimension is eroded in another. The machine, which was to be a servant, is more like a master. Or rather, as we have attempted to absorb the machine, it has absorbed us as parts in its functioning, permitting small tolerance in its demanding efficiencics.We have been brought to subservience by force of competition, and are spurred by the terror of replacement which must be the dread of every spare part—a dread which must increase precariously as unbridled population increase demolishes the ground of individual right and dignity. Again, in the efforts to bring abundance to destitute populations, any gains are rapidly diluted by numerical increase. Such are the crudest indications of tantalizing defects, sustained by both rich and poor. As for disarmament, it is impossible in the present order. The world disarmed would be like a knife standing on its edge: the slightest breath would make it fall over.

With such insoluble problems of need, the only course is not to hold them inch by inch as they force the pace, but to cut through to a position in which they are not resolved, but dissolved. A definite position which can be held is that restraint on parenthood, as a universal principle, is the keystone for an order in which these basic problems can be dissolved; moreover, an order in which liberty would have a new amplitude of realizatron, and some notion of brotherhood, which is essential to man, could be more than an incanted phrase. It could become a general essence in conduct, an influence which would elicit that more agreeable nature which is potential in humanity, and allow to fade those familiar barbarities which, perhaps splendid in another setting, in the circumstances and possibilities of the age are only a tedium. The destructive bomb, and the finiteness of the earth, have been in the historic distance, a threshold which must sooner or later be approached. But that threshold has now come into view, bringing the moment of truth. It is time to regather an obsolete order of things into a new shape, not by violence, but, on the contrary, by an unexcessive restraint whose burden and great advantage all would share.

This restraint cannot be achieved merely by pointing to alarming statistics and spreading news about techniques of birth control. People do not seem to concern themselves much with a distant threat, particularly if it is universally shared. Or, if they do, the threat is allayed by some distant hope—though in this case if it is not an absurd hope like colonization of the planets, it is hard to find. The restraint has to be instituted within an encompassing perspective—the hardening of which is the essential endeavor—and by a definite principle the necessity of which is abundantly clear, as is generally the case with murder and theft. Such a principle is available: it is merely that to replace oneself is a right, a liberty without infringement; but to exceed replacement is a crime, a failing in regard to mankind as a whole, having the nature of an aggression. which, if universally practiced (like the burning of coal fires in fog-bound London), brings universal loss. To gloss over simple difficulties in human arithmetic, parenthood by each individual of two children, on the average, is replacement; but parenthood of three amounts to a rapid increase, which will eventually have to be curtailed by destruction. There appears to be a choice: are children to be brought into the world, haphazardly, to contend with each other for a place to live, and be brutalized by crude struggle and valueless demands to the extinction of innocent and valuable aspirations, and in the mass often to be regimented for violence; or are they to enter in the liberty of acceptancein a kind of general family? The earth needs to be cultivated rather than consumed, and there is only dissolution in the haste and pressure of population advance.

Such a principle of restrained parenthood has as one important corollary the containment of population. It would not be a revolution in the sense of direct violence done to institutions. Nevertheless it would have revolutionary effects. But, besides the most tangible realities, such as make for popular anxiety about the "explosion" of population, or of bombs, it would touch far reaching complexes of change which would be more in the sphere of evolution than of revolution.

Both explosions, of population and of bombs, in any case facets in a general explosion of imperfectly assimilated knowledge, give a good forward perspective.Without them, the age-old floundering between atrocities in war and extremities of tyranny and injustice in peace would, it must seem, drag on endlessly, and be accepted as part of a "natural" order—however fine may become the knowledge of order, as such. But that drag now has a visible end which must bring about a certain arrest. At least the turbulence may be elevated to a new sphere, which can be the stage for all kinds of drama, but without the likelihood of bringing about the dissolution or the total destruction of all inheritance. Though he will never become peaceful, the military pacification of man is bound to become a commonplace. Swords, no doubt, will be transmuted into scientific axes and metaphysical thorns, and applied to terrible argument.

Capability for war is itself in process of forcing military peace, in some form or other; and the process must certainly be completed, because the thermonuclear bomb, whether or not it is usable, is not the end of military technology. But military peace is no culmination of value. With life, always thrown on as it is, ever out of equilibrium, there must always be some kind of assault; and the enthusiasm for it is inborn in man. But the generalization of struggle and assault, beyond mere physical assertion, is a peculiar attribute of man, a great distinction, which can even be an essential definition. However, the old, now abortive modes of aggression could not be abandoned voluntarily, but from an extremity of necessity such as now exists. The bird did not take to flight out of love of the air, of which it had slight experience, but out of a fearful respect for the ground.

Economic need is often considered primary. In a mere quantitative aspect it provides a basic motive, the manipulation of which is a coercive measure inferior only to military force. With a stable population, coupled with the automation of labor, every kind of economic provision would come so amply within the bounds of resource and technique that the quantitative factor of need must dwindle to insignificance. The slackened traces of brute economic necessity would leave further pursuits unencumbered. The paradox that scientific advance has created conditions for a new bondage, instead of realizing the hope of a new vista for freedom and growth, could thus have a chance of being resolved; and also perhaps one of the main points for the gathering of tensions would have been removed.

From population pressure many other pressures stem. Obviously there are pressures on every kind of limited resource. From their intensification and the heightening of the scramble after needs, there results a destabilization which will bring about crises of aggression. Population pressure can be the basis of imminent violence and cruelty, outwardly in the form of war, and inwardly in the texture of the life of the society. But in addition to the very tangible effects of population increase, such as a loss of physical space and resource, and perhaps even dignity, there are other effects which are just as important but more difficult to identify. It could be suggested, for instance, that there is a loss of psychological space. There is an invisible but heavy hand which brings about a psychological confinement. To elaborate on such a notion is important, but the propositions that can be found are, by their nature, only more speculative. However, the peculiar tyrannies that can be born within teeming populations must be at the heart of the interest in the "explosion".

One fundamental handicap of an exploding population is that the processes of education as we know them now will not suffice. Intensive, massive competition among unlimited candidates for limited opportunities makes for a narrowing of margins, and subordination to an ever more limited expediency. Education has become something like a military training program in preparation for an assault, which, once the strategic position is gained, can be abandoned. It is destroyed as a value continuous with the general on-going of life. But mental life is the one distinctive human attribute. It carries within it the source, and even the essential meaning, of liberty. To limit its cultivation to the gross functions imposed by competition is a significant step towards the atrophy which is the condition for tyranny. To make for liberty, a sense of play must be given rein and allowed to grow. It is the sense of play, and the capacity for pleasure, that is the index of culture in a community. Play is at the bottom of growth and originality; it is in the mode of creation. The vast multiplication of like things, as in industry, or college, or population, is not the same as creation. Perhaps in a future age, assuming that mankind emerges from the crisis of the present time, the mania for material production which is a modern predicament will be looked upon as a crippling disorder, just as we now look upon the frenzies of other ages past. Demands on energy and every resource need to be kept within bounds, lest in the overwhelming tension, gross expediency will gain the upper hand and with continual weakening the whole familiar edifice of value will drift away in a deluge of millions and billions of people, to leave a process of life which could hardly be called human. When Noah builds his ark again, perhaps he will be less scrupulous in taking his passengers two by two.

An actor needs a stage, a fluid has its vessel, space requires coordinates, and so forth. Nothing exists without its form. To have form, and therefore to be limited, is the price of existence. The extremity of liberty is not to exist at all. But here it is no fine philosophical matter, but an urgent choice. The preachers of nothing but liberty need to be more explicit. To many, an absolute restraint on parenthood, in the sense of a right to replication, but not to multiplication, will seem intolerable. But it can be contended that now in the scales of choice is the private freedom of parenthood against the very existence of the race. This is a contention which can call for much exploration, which can be strongly defended, and which can also always be doubted. In fact, it would seem to be a dilemma. But every child, it has to be seen, is also the world's child; for two parents to have two offspring should make for no privation, if that is the case.

ABSTRACT

AFRIAT (S. N.) "People and Population." <u>World Politics</u> 17 (3),
April 1965.
Japanese translation with foreword by Edwin O. Reischauer
(US Ambassador to Japan):
<u>Japan-America Forum</u> 11, 10 (October 1965), 1-28:

The explosive growth of population is the central
reality in every perspective on the future. But, just as the prize for a
successful confrontation of the existence of the thermonuclear
bomb is military peace - a gift which, experience suggests,
could not be given otherwise - so also the looming population
crisis is a spur to desired ends which could not otherwise he
approached . The confrontation of the problem is not only
essential to the future, but is the keystone of an order in which there
can be dissolution of problems which are now critical but, in the present
order, seem insoluble. It is a moral issue as fundamental and far-reaching as
curtailment of theft and murder. Whatever the possible forms for its
realization, the principle in process of emergence is that, while
self-replacement is the right of an individual, to exceed
replacement has the nature of an aggression against society and against
mankind as a whole.

(International Political Science Abstracts, UNESCO)

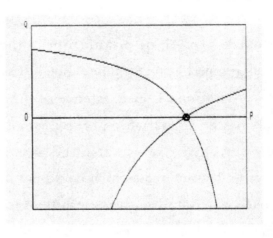

4
Market & Myth

1994

In speech of today a single being, the Market, is at the centre of existence. It is the self-created, self-regulated, unquestioned measure of everything. The textbooks teach that, but for objectionable intrusions, or 'imperfections', its rule would be *optimal*.

It has been remarked that every age has its myth proclaimed as the higher truth; here we have ours. It permeates every sphere—which could perhaps interfere with visibility. As a movement in the immediate present it is on a gallop with fresh energy commanding submission everywhere.

In history this myth, and the reality which would go with it where the human factor is something like an accident, is rather new. Now it is propagated by endlessly repeated teaching, and bolstered by interests, and inertia. A first impulse for present objections came simply from encounter with the textbook teaching, and being struck by its thorough absurdity and deceptive use of language. This is unaltered by the mathematical dress given to it, and enunciation of portentous theorems. While such presentation may give it all an aura of lofty science and unassailable truth, it is anything but that.

Such a concern might appear simply scholastic, a field of sport for any with the training and inclination and otherwise of no account, like many another. Faults with the teaching have been pointed out many times, but this appears to make no difference. As indoctrination it has a now ominous persistence. With inheritance from the 19th century based on inchoate notions from France in the 17th, we are approaching the 21st. Instead of being rudely awakened, as happened lately on another side, it could still be asked whether this is the way to face the now well recognized and already well advanced, developing and encircling realities.

The question requires a regard for the market teaching which has importance from its influence, and apart from that is an exhibition of self-deception—a spectacle to make the power of reason as much of a myth as what it pretends to offer. Of course we know reason has its intermittences; and F. A. Hayek has spoken well about authority in economics and the transmission of mistakes, how they are handed down with uncritical acceptance simply because of the prestige of their perpetrators.

Without the barriers that not long ago always prevailed, when markets were auxiliaries, subordinate to life in a locality, the market dictatorship would most likely impose itself anyway, by its own law, regardless of what anyone should say or think about it. As well accepted, liberty requires a constant defence. Now the absolute rule is fostered by absurd reasoning of scholastics, joined with zealous belief of creatures of the days of innocence which are no longer with us.

We have a concern with the Maximum Doctrine of Perfect Competition which is central to the dominant 'neoclassical' economics. It appears to have originated in the seventeenth century with François Quesnay and the Physiocrats, and was imported from France into the United States by Dupont. For perfect competition there are the well known conditions, and the conclusion—which for the Physiocrats was nothing but self-evident—is that under these conditions the economy achieves an optimum. This is the "social maximum" alluded to by Kenneth J. Arrow in motivating his theory of *Social Choice and Individual Values*:

> If we continue the traditional identification of rationality with a maximization of some sort, then the problem of achieving a social maximum derived from individual desires is precisely the problem which has been central to the field of welfare economics.

What is the criterion by which one can know the better and the worse and hence what is best, or the optimum? Nobody knows. Without a knowledge of this what we have been told must be quite empty. There could be cause for an abandonment of the whole idea, and surprise at how it is accepted without a tremor by the multitude of the faithful.

To be *free, and yet a good slave*—put that way it sounds ridiculous, though it should strike one the teaching is just like that. First there is the individual freedom in the self-regulated order, the market. Then as if here is not enough to the system, and in further praise of it, it is submitted that the overall result is *efficient*, as an obedient slave performing some precise duty to the utmost. It is a relief one never is told what the duty is. The social objective it is taken to exist and to govern because it is talked about. With more known about it there would be a better position to verify whether or not it is at a maximum. For some the loose end is put out of the way as the Aggregation Problem, but should anyone ever get to that problem they would not know what it is.

We are faced with a phenomenon appreciated in another case, the famous "happiness" formula, known as a Marxist slogan though it has an earlier origin. P. P. Wiener attributes it to Francis Hutchinson the teacher of Adam Smith. Its classic attribution is to the Utilitarians and Marxists must have borrowed it from them. According to I. Philips:

> John Bowring says in his Deontology [1834, p.100] that Jeremy Bentham recalled how on a visit to Oxford in 1768 he had first come across the phrase 'the greatest happiness of the greatest number', in Joseph Priestley's Essay on the first principles of Government, published in that year, 1768. "It was from that pamphlet [Bentham said] … that I drew the phrase, the words and import of which have been so widely diffused over the civilized world. At the sight of it, I cried out, as it were in an inward ecstasy like Archimedes on the discovery of the fundamental principle of hydrostatics, $E \upsilon \rho \varepsilon \kappa \alpha$."

We should try to find out what the stirring formula could possibly mean. Since "widely diffused" without any qualification, we may look for its import in a simple possible world, one where a cake is distributed over a number n and happiness is the size h of the slice anyone gets. Then the greatest h for the greatest n is wanted. To put all this mathematically, with the size of a slice measured by its angle in radians, so the whole cake is 2π we have the constraint $hn \le 2\pi$ and have to maximize h and n simultaneously. Let anyone try!

Economics students receive the notion that if no one can have more, unless someone has less, then we have an "optimum". It is tagged with Pareto's name. It is just like with the cake, so apparently you can distribute it around to everyone any way you please, it's always optimal. Good news for the party host as for the economics catechism. Since everyone wants more, this would have to be a case of "Multi-objective Optimization"—the title of a lecture I once saw announced. But there can be no such thing. If you have one objective then you cannot at the same time also have another—you just have to make up your mind!

Impressive absurdities on the same model had occurred previously, for instance Quesnay's Economic Principle "greatest satisfaction with the least labour-pain", and he must have drawn inspiration from Leibniz whose "best of all possible worlds" provided the greatest good at the cost of the least evil. The precedents give a reminder of Hayek's remark about the transmission of mistakes.

Obviously if you choose to maximize one thing, then you cannot at the same time make a free choice of another. You may be lucky, for instance if (x, y) is subject to $x \le 1, y \le 1$ and you want to simultaneously maximize x and y, this is provided by $(1,1)$. But we do not have a case like this in dealing with the "happiness" formula, or the cake; for when n is made large h is forced to be small, and *vice versa*.

It may be wondered how anyone, whose respected output is supposed to be rational (in an ordinary sense), can make such remarks, and how they can then have acceptance, even be awarded prizes. On submitting about wrong reasons to Chalongphob Sussangkarn, on a visit with a Thai trade delegation, he gave a healthy answer: "We have the right thing—never mind those reasons!"

Here is another thought bright with free market devotion, from Robert Heilbroner in *The Worldly Philosophers*:

> Edgeworth's pleasure machine assumption bore wonderful intellectual fruit ... it could be shown—with all the irrefutability of the differential

calculus—that in a world of perfect competition each pleasure machine would achieve the highest amount of pleasure that could be meted out by society.

Enjoyment of the wonderful fruit should in this case be spoilt by a suspicion of worms. What is "all the irrefutability of the differential calculus"? Is it irresistible authority of the Chain Rule; or final truth in the Infinitesimal unphased by digital diversions; or the incomprehension and boredom of all those readers who give a passing glance at the exhibition of machinery and then get on with the text? We should do that first since the outer skin of this fruit is not without blemishes. We are faced once more with the Leibnizian nonsense, expanded into n dimensions. That ought to be a relief since now there should really be no need to go back to that skipped-over calculus after all. However, belief that there is complete relief is feeble *optimism*—a dream of *rationality*. The particular calculus turns up in countless textbooks—at least now we may perhaps know where it started.

For a separate matter where there is a striking inconsistency, something like in the "happiness" formula, Mr James Baker the erstwhile U.S.Secretary of State toured the newly independent republics of Central Asia, speaking with their leaders and submitting what is expected of them: "democratic government and free-market economics". The principle of such government must include some independence. In allowing that, how can it also be laid down what they should decide? A people could wish to maintain competition internally to brace up performance and exploit local capacities, perhaps on the side of basics for people and territory, without putting these at the mercy of a noisy global competition for which they are thoroughly ill-prepared—in other words, settle for living happily with their comparative disadvantage. Instead of doubling their population in thirty years or so they might even choose to limit themselves—and pursue "greatest happiness" for their steady number! After all, if one couple have three children its an appalling 50% expansion in one generation, draining away surplus for improvements, if any. Human rights which get eager attention and have been listed at Helsinki are no doubt good. Nonetheless it is not always clear where the rights come from, and whether people in a chaotically crowded world have any rights at all. How about obligations, and shouldn't they come first? The signatories of the accord could then form a truly well-considered trading block.

There is a strong tide in favour of knocking down trade barriers and fostering growth, but also a rising opposition coming out of concerns with ecology, resources, demographics, environment and the like, and deteriorations throughout the world. At an earlier time the collision was rather between economic and social factors. Karl Polanyi (1944) gives an account. Now a completely new era has arrived in which hanging over everything are broad questions of survival with a rather short time horizon. Near the threshold of subsistence there are no margins, and with a tight global competition, in a great part divorced from local attachments, there are none either. That covers the most numerous and also the most rich and powerful elements. So where is the will or strength to alter the current course? This is the question dealt with provocatively by Richard Falk in his *Explorations at the Edge of Time*.

J. M. Keynes reviewed the accounts and gave a reconsideration of usual wisdom about trade in his 1933 article on "National Self-sufficiency". He was perhaps thinking about other things, but what he had to say may cut better now than it could at that time.

The Earth is not a remote abstraction but a patchwork of localities, each quite immediate to whoever happens to be there. Nor is humanity, which consists of actual people all living somewhere. Beside the currently topical global outlook, if there is a relationship that has priority it is that between people and where they live. The place is a first base for life and sustenance of inhabitants, who are its custodians; all that goes with it is their care and responsibility, and if they make a disaster of it it is mainly their hardship and their own fault. Such a severe view seems in harmony with remarks of Keynes, and invites consideration as a response to ruinations that take place. In any case, there is unrest in being dependent for vital needs on trade with others with whom no steadier bond exists.

Since every era has something like its own economics, dissatisfaction with what we now have may reflect a transition. But to what? Some say a "new paradigm" without suggestion of its nature or what should be done about it. Propensities in the neoclassical and Marxist phases give visions of another heady round of 'theory'. The neoclassical outlook originated from the time of Newton and the euphoria over his mechanics, and the 'Optimism' of Leibniz. The economy had then to be approached as a machine, not well understood because nobody around had made it or had the plan. Hence the models economists play with, and the cult of the optimum—"the best is enemy of the good" may be recalled at this point. Marx under the spell of fashionable Hegel took on his notions helplessly, and adopted Ricardo's theory of value without noticing the actual arithmetic is impossible. What could be next? Documentation of happenings to the globe such as State of the World reports of the Worldwatch Institute, or *The Ecologist*, suggest things are going to be different. They cannot be the better—let alone *optimal*, if that myth can be dispelled—simply by submitting to government by the self-regulating machine and giving anxious attention to its ups and downs that dominate the news. There must no doubt be withdrawal from that foolishness, though there cannot be a ready-made design for an era with unprecedented features—and touch of finality.

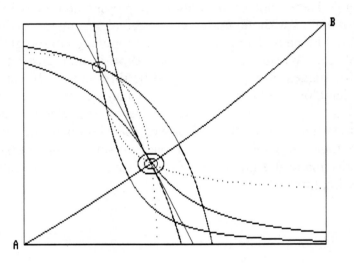

5
On Trade,
and Self-sufficiency

Symposium on
Equity and Efficiency in Economic Development
in Honour of Benjamin Higgins
1991

1 An Introduction to Economics

I welcome this opportunity to pay tribute to Benjamin Higgins, a few recollections will not be out of place, and they provide a setting for what I submit on questions we have sometimes tried to discuss. Our paths crossed at various times since the 1960s. We first met when he was at University of Texas in Austin and I at Rice University, Houston. Then in Ottawa I joined him as a colleague. His retirement—not the suitable word nor is 'advancement' though 'advance' might do—was a loss. He went to Australia, homeland of Jean Higgins. Meetings did not so much involve economics as conviviality. When we did get to economics the recurrent theme was, as now, that something was wrong. Aware of a differences in focus and failure of language, we never got anywhere and left the subject, only to return. Perhaps it was disposition as writers instead of readers, or listeners, that impeded communication. This occasion gives a chance to bridge the gap.

Visiting Australia in 1987 I spent the first weekend at the Higgins's 'Post' (a thousand acres, some sheep). After misadventure with the train from Canberra, Jean Higgins turned up in Cooma to drive us to Kallarroo. In that incredible isolation before a blazing fire fed from gum trees that, because of compensating drafts, did little to warm the Post, the subject came up again.

For Thomas Malthus it may have been the "dismal science" but of course economics is not free of strong feelings nor a science. For lack of a common bedrock positions are governed by variable opinion, tradition, pressure, personal belief or interest, or some respected "scribbler". An amount of heat goes into whether markets should be 'free' or not. Credibility is a constant issue—present company not excepted.

Graeme Dorrance from Australian National University was among the group assembled there. He said with quiet confidence of settled belief that it was odd to have complaint from one such as myself, a typical embodiment of what was wrong with the profession. My association is with mathematical economics and he shared the idea that salvation would come only when such diversions are put aside to allow a genuine concern with real issues.

I would respond by saying that closeness to issues may give relevance and weight and a good conscience but must be a distraction. As it was with another doctrine lately, so it is with economic teaching: while certain things may have important influence they are nevertheless misleading, an abuse of mind and language. Should economists have a concern with such issues? In the main the profession continues regardless, with little self-examination. Government of the group decides the games to play and the "Crisis in Economic Theory" which has been so much discussed lately is a tantrum in the Toy Department.

An interesting line to pursue in trying to understand something of the sort that may be wrong is the "Optimism" associated with the idea that trade is connected with a "social maximum"—whatever that may mean. If this were excluded from belief some usual pieties about trade would not look so good and one might even turn to common sense.

At the World Congress of the Econometric Society in Cambridge–UK after he had won the Nobel Memorial Prize, Paul Samuelson confessed his motivation: "adulation of the economics profession". As much as any reality to do with economy, also the profession itself is worthy of study. We already have Axel Leijonhufvud's classic on the model of Gulliver[16]. Another contribution comes from Robert Kuttner[17]. At the start Kuttner says "Events have been unkind to the economy, and unkinder still to economists," and "Since 1970 an outpouring of serious and ideologically diverse articles and books has pronounced that economics is in a state of severe, perhaps terminal, crisis." Vincent J. Tarascio writes[18]: "The purpose of this paper is to examine the nature of this crisis from the point of view of the sociology of our discipline." There seems to be a sharper focus on life of the profession but "the ship

---

[16] "Life Among the Econ", *Western Economic Journal* 11, 3, September 1973. Reprinted in Dimand (1986).
[17] "The Poverty of Economics: a report on a discipline riven with epistemological doubt on the one hand and rigid formalism on the other", *Atlantic Monthly* February 1985.
[18] "The Crisis in Economic Theory: a Sociological Perspective", *Research in the History of Economic Thought and Methodology*, 4, 1986, 283-95.

sails on" and what difference will it make?[19] Thomas Malthus was worried about abuse of language in his *Definitions in Political Economy* (1827) and he would not be happier now.

Clearing decks helps with management of disorder and creates space. Certainly the "Optimism" should be given up and the wilderness that results would have to be an improvement.

When a distinguished practical capitalist can express in public his reservations about "free-market" culture, it makes news.[20] For now we have from George Soros[21]: "Although I have made a fortune in the financial markets, I now fear that the untrammelled intensification of laissez-faire capitalism and the spread of market values into all areas of life is endangering our open and democratic society. The main enemy of the open society, I believe, is no longer the communist but the capitalist threat." But where is—or was, when it was first heard about from Karl Popper—the "open society"?

The abandonment of structures leads back towards a primitive level, even to Aristotle. As a complement to Keynes[22], Aristotle's views about trade and self-sufficiency are impressive. He requires that trade be in the service of self-sufficiency and should not go further (*Politics*, Book I). Adam Smith notwithstanding, the meaning behind this idea may be as worthy of study now as it might have been then.

No one can dispute that trade is basic to economic life even when absurdities about its merits are put aside. At the same time there is no simple notion of what should be meant by self-sufficiency. Opposing views come forward in times of change, so it should not be surprising to hear in a given moment that "In this day and age, there is no such thing as economic self-sufficiency" (Henri de Villiers quoted in *Time*, March

---

[19] "The Decline of Economics." *New Yorker*, December 2, 1996, 50-60.

[20] Eric Ipsen, "Capitalism's King Spies Evil in Market-Mad Realm." *International Herald Tribune*, January 16, 1997, pp 1 & 6.

[21] "The Capitalist Threat." Atlantic Monthly, February 1997, 45-58.

[22] J. M. Keynes, "National Self-Sufficiency", New Statesman and Nation, 8 and 15 July 1933; Yale Review, Summer 1933; Collected Writings Vol. XXI, 233-46. I am indebted to Jim Alvey, of Macquarie University, NSW, for drawing my attention to this paper.

27, 1989, p. 42) and "There's no longer any such thing as state sovereignty" (Allan Gotlieb quoted in *Ottawa Magazine*, March 1989, p. 22). Ages pass and this one may do so as promptly as those of recent memory when it was confidently assumed the Affluent Society was here to stay, or that communism would take over. Another has arrived already, when the Earth the source of maintenance has maintenance problems itself.

Conflicts between major powers must in some manner be receding. Size served power, but if the power itself becomes obsolete so perhaps does the size. Then there may be a yielding to the smaller groups clamouring for separation. These peoples can then properly look after themselves, stimulated by the immediacy of effects of self-management, on a scale where damage would be more local than global. Maintenance of the Earth may then be more generously forthcoming, since coercion that springs from directly felt self-interest is the most acceptable kind and most to be trusted.

In any case, there might be room for speculations over what remains but none over the part under the "Optimism" heading central to the present exploration, related as it is to the inchoate intellectual fumblings of early days that persist in the textbooks and minds of well-trained economists. This is the main of what is offered here, the remainder rather having the nature of an application exercise or loose enquiry. Unburdened of such futilities there is more room to expand, even freedom to arrive—if anyone should insist on some outright simple concept of 'self-sufficiency'—at an absurdity, just the obverse of the one we left.

Kenneth Boulding was able to touch on "The Legitimation of the Market"[23] but legitimacy for the pervasive common-sense practice of economic protection is less readily granted. It is kept that way by what amounts to a metaphysical belief, that an overriding objective should be to knock down barriers to trade to allow domination of a global competition serving an unknown end. Instead of this determination, another objective has been coming forward, one that gives another

---

[23] Lecture at Rice University, Houston, Texas, 1963.

dimension to efficiency and seems to have more to do with cooperation than competition, namely, the objective of survival.

An efficiency for the market may be granted; there could be a sense to that somewhere though it is certainly not where it is usually put; and in any case markets always spring up when allowed. But such market efficiency could not then be an efficiency as commonly understood. If not just because of the old mistake, perhaps also from a disinclination to capture this sort of efficiency for what it is, economists have come to pick on another efficiency that, to the extent it is in any way understandable, lies altogether in the realm of myth.

The record of failure of development programs over some decades is frequently noted. Benjamin Higgins, seeking a definition of development, names six proposed definitions and asks "How can reasonable men reach such diverse conclusions?"[24] Another definition may be considered in different ways, both reasonable and unreasonable: the building of self-sufficiency. Such an idea may go against formulas and interests but it has a practical even if shadowy presence. A part of the inclination towards sovereignty is for some form of self-sufficiency. Besides the geographic boundaries, lines are drawn about what is for sale or to be bought, and who are the partners. Inevitably, or for good reason, important capacities are not allowed to atrophy and wither through the exploitation of markets and dependence on vagaries of others.

In regard to economics, and whatever else, with changes taking place a framework is becoming settled where modes of thinking have to be different. Discarding old and taking on new is not a textbook matter such as might be assigned to the care of some accredited professional. Part of the process is liberation from baffled preoccupations of scholastics as well as from absurdities found therein.

Questions about human life have recently come to a new point, where the Earth, once taken for granted as the ready source of all sustenance, has itself become an issue. Aside from the demands of

---

[24] "Equity and Efficiency in Development: Basic Concepts", in Brecher and Savoie (1993), Chapter 1.

increase of populations, the global economy is a machine for destruction of the Earth itself as well as less earthy inheritance. For a liveable future there has to be a escape from this domination. That this is so is largely left unsaid in declarations about the environment that have been heard everywhere recently.

In the 1960s there was an outbreak of concern about "the population explosion". In 1965 the problem was recognized for the first time at the level of governments, so proponents of population restraint marked that year as the "year of the breakthrough". However, the less-developed countries had this to say: "Make us rich like yourselves and then we will have fewer children." After that the whole matter seemed to disappear underground.

Although public attention has drifted away from the overpopulation issue it remains fundamental. It is commonly thought of as a modern problem noted by Thomas Malthus, but the following gives another idea:

> There was a time when numberless races of men wandered the earth Seeing this Zeus took pity and resolved in the wisdom of his heart to relieve the all- nourishing Earth of men, stirring up the great quarrel of the Trojan war in order to lighten the burden by death. The heroes perished in Troy and Zeus' plan succeeded.

The Cypria, attributed to **Stasinos**, *ca.* 7-5th century BC

In recent times there has been a speeding up of the exhaustion process that has been going on for centuries. Fertile regions that once supplied the granaries in Rome and others exploited by ancient people are now deserts. Vast forests disappeared simply to supply wood to build those great fleets that used to sail. We already have an impoverished Earth. In days gone by there was always somewhere else to spoil. New worlds are no longer so rich, and depredations that formerly might have taken centuries can now be accomplished quickly.

Destructiveness of warfare has in some way abated, not through any new wisdom but as a blessing of the Bomb. There is no economic bomb as yet to bring about a pause; rather, we remain creatures and prisoners of a system that knows only itself and nothing of its results.

In the 1970s we heard news about "Limits to Growth" supported by computer printouts and ominous pronouncements from the Club of Rome. It is forgotten now, a fanciful moment swept away by urgencies of the real business of living: market competition and growth. The 1980s saw the globalization of markets and rise of Japan as an economic power. The liberalization of trade and the need to be competitive in the world market are proclaimed as ruling principles. Surviving in this economic reality makes an intimidating prospect, even for the powerful.

At the start of the 1990s trading blocks were taking shape, lines being drawn as though for battle. While a withdrawal from trade may not now be occasion for visit of a gunboat, multilateral liberalization is dominant, while voices express concern at signs of a rise in bilateralism, at sovereign parties independently getting together to serve their own separate interests.

The failure of communism may be a vindication of the market principle, among other things, but seems to have occurred at a problematic moment for capitalism, where beside debt there is the many-sided problem related to environment. The response to this has included measures often to be brought in gradually, or just studied. For with the preoccupation with competition, and debt, the problem cannot receive due attention. There is hardly a suggestion that current economic thinking should change drastically, let alone that it may be destined for anything like the fate suffered by communism. Even the search for a 'New Economic Order' which had prominence during the 1960s and 1970s had no manifest outcome.

Many concerned with ecology, environment, and population, remain pessimistic, unsure that a momentum that has grown out of the entirety of past history can be deflected to deal with an unprecedented and abruptly arrived global problem. Evidence for looming disaster is ample enough, whether it should issue from scientific studies or common sense. But such evidence appeals to reason, a poor teacher and mover to contend with overreaching continuities; another teacher, if there should be need of one, would be breakdown and calamity.

While markets and trade always remain central to economics, teachings spoil this matter with absurd simplistics and sublime myths inviting the abolition of all barriers—as it were putting everything for sale. There has no doubt been a great trading era, underpinned by devotion to its own myth, the market. One could say: "Never mind wrong reasons, we have the right thing"[25]. But those reasons, spurious though they be, have a sufficient durability in minds and textbooks to give sanction to an order that is outliving its welcome.

2 The 'Optimism' of Market Doctrine[26]

2.1 Choice and welfare

The thread taken up in this essay has connections with 'choice theory' which, being a subject peculiar and distinct, warrants separate attention. Choice theory has a relevance to development interests and may be associated with efficiency taken broadly. As for equity, F. A. Hayek[27] has provided an interesting view. A similar, though thoroughly antique, notion is 'fair trade'. With all the modern market wisdom however, fair trade still finds its way into thinking that goes on widely and at every level, though no one quite knows what it means. It seems to be a survivor from early societies, maybe properly active in some situations, where, though it is unspoken, everyone knows what is due to them, or is fair. With trade agreements, however, except where fairness is reflected in some clearly understood reciprocity arrangement, who knows if they are fair or not? Apparently, fairness can only be measured by the degree

---

[25] When I submitted these ideas about wrong reasons to Chalongphob Sussangkarn on a visit with a Thai trade delegation he replied (I thought very well) with a remark something like this.

[26] Based on my *Logic of Choice and Economic Theory*, OUP 1987, Sec. I-18 on "The Maximum Doctrine".

[27] "The Atavism of Social Justice", Nineth R. C. Mills Memorial Lecture, University of Sydney, 6 October 1976. In Hayek (1978), Chap. 5.

to which parties honour the agreement, and there certainly can be disputes about that.

Making choices is important for economics—all that could be more important is having the opportunity. But it makes an unresolved subject, even where the issues touched are basic. Formal choice theory seems peculiar to economics and it started early. Joseph A. Schumpeter[28] attributes the "Economic Principle", which joins with the idea that the economic problem is a maximum problem, to Fran◊ois Quesnay (1694-1774). It remains permeating the subject as much as ever and a question is whether it overran its proper course. Probably it did that in the beginning and the early words have had a survival.

The accumulation of attention is in proportion to the duration, and though there has been fascinated attention to the works of the clock still we are not sure of the time. Matters there are due for settlement, for we have been told "fools rush in where angels fear to tread" and in having that there would be better chance of company with the latter. The former might make discoveries but that is not quite what is wanted, though one could be without any and still be far from the latter. Elaborate structures have been built on precarious drifts of meaning; there might be a desert without them—but then getting used to it would be much more economical!

A cause for some general confusion is ambiguity. What is meant by a choice may be clear, but how, where, or why a choice should be perceived in the first place may often be less clear. There are problems with terms like 'preference', 'optimum', 'efficient' and 'welfare' for an individual, a group, or an economy. Often one might wonder whether some proposition is true or false, or neither. Different organizations or disorganizations of concepts have simultaneous use and following ordinary usage with key words would be helpful. For example generalized preferences, without the usual transitivity[29], are certainly strange; once it is possible to talk about such things an anchor has gone and anything can be called anything.

---

[28] History of Economic Analysis, Oxford University Press, NY 1954.
[29] If A is better than B, and B is better than C, then (for most of us) A must be better than C.

49

A similar case is the "Pareto optimum". There can be dissatisfaction about a doctrine that has early origins but still prevails and is represented in the textbooks. A reading of the "Maximum Doctrine" of the Physiocrats, which is meaningless taken literally, has been translated into a misreading of Adam Smith's doctrine of the "Invisible Hand", and this in the hands of mathematical economists using set language and the like has been translated again, but not very well. In the latest version we have the Pareto optimum. When that is seen for what it means, in no ordinary sense is it an optimum: it is just called that while the power it has in economic thinking is as if it were that. Pareto fleetingly entertained the idea as being analogous to a maximum and it has come to have exaggerated importance. It just filled the vacuum created by the shortage of meaning in the old doctrine.

Even if we are assured that Adam Smith did propose a maximality under government by the Invisible Hand—and it is quite possible that he did[30]—we still should not take the idea seriously. It could be a quaint residue of early thought—after all, Newton's mechanics is not vitiated by the importance he gave to number magic and alchemy (perhaps the contrary now, but we can put that aside). It does not matter what views the Physiocrats or others had about automatic global economic optimization under various conditions that can be spelt out carefully at length, we still should not believe in them for we do not and cannot possibly know what they mean.

To the Physiocrats the Maximum Doctrine was not a matter requiring proof—it was self-evident! There have been gestures to prove it since, out of respect for the old words mixed with duty to contemporary science, but no one knew quite what it was that should be proved. Words have patterns both with and apart from their meaning, as recognized in songs. As interesting as this matter itself is the way it has

[30] Tom Settle, Guelph University, has assured me that he did, and provided a copy of the relevant passage which unfortunately I have lost. Settle (1976) forcefully expresses a view similar to the one set out here of the usual textbook teaching.

been preserved, and conditions thinking still. This phenomenon of transmission of authority was pointed out by F.A. Hayek[31] .

2.2 Free and yet a good slave—or Optimism

This title may not be from a well-used stock so that it should tell plainly what it is about, but it suits its purpose. This is pursuit of a thread—that makes embroidery with the optimum, competition, efficiency, welfare and the like—that has run through discourse from early times into the present and latest textbooks. There is an endless repetition and we want to find out what to make of it. If there should be something wrong we would still like to understand whether it is good or not—to know, so to speak, the welfare of it.

What is submitted here has been offered by many writers each in their own way, but this appears to make no difference. How then should one deal with the matter? Perhaps with humour, and a look at history. It is not important whether everything reported be right or wrong, as long as things are presented in the right light. For what we have to consider has its own evidence which has nothing to do with history. One may take the clear path of simply looking at the matter itself. But the popularity faced is resistant, and for the reason here the early story may have revelations; in any case, a glance at the salient can come first.

The association of general economic equilibrium, on some model, with a social optimum, or maximum, is paramount in economic teaching. This was the start of welfare economics and related free trade polemics. There can be an approach where everything is reviewed from the ground up, and another, as it were the contrapositive, where we look first at the fruit. The latter is least laborious and enough to raise questions.

To be *free, and yet a good slave*—put that way it seems ridiculous, but it should strike one that the teaching is just like that. First there is the individual freedom, in the self-created, self-regulated, stable order,

---

[31] "The Pretence of Knowledge", Nobel Memorial Lecture, Stockholm, 11 December 1974. In Hayek (1978), Chap. 2.

the market. Then as if this were not enough to the system, and in further praise of it, it is submitted that the overall result is efficient, like an obedient slave performing some precise duty to the utmost.

It is a relief that one is never told what the duty is. The social objective is taken to exist and to govern—because it is talked about—and there is discourse on properties of the 'social welfare function'—they are 'revealed'! If more were known about the welfare function there would be in a better position to verify whether or not it is at a maximum. In some minds the loose end is put out of the way by a transfer to the Aggregation Problem, but should anyone ever get to that problem they would not know what it is.

The efficiency entertained in this story is based on the 'commodity space', or some derivative—in its earliest instance quite nebulous and later involving utility. The free market may truly have a genuine efficiency, of some sort, but then it would be in another space, not one that has a part in the model. Perhaps it may in some way have to do with taking over the otherwise formidable task of coordinating supply and demand—to encourage some fair allowance for the story, if that were to be an objective.

In one form of the doctrine competition is central, complex, and carefully spelt out at length. In a later form it turns up just as a word, tacked onto statements but doing no work. Here one might puzzle over the real importance of competition and 'competitive equilibrium'. On the other hand, we do recognize the value of competition and its results even if it cannot necessarily be captured in a model. Competition is a stimulus with unheard of results, and having anything unheard of represented in a model amounts to a contradiction in terms.

It takes only two to make a competition more bedrock than perfect competition. There is also useless competition. Competition is ordinarily understood not to be unbridled but to be confined to limited channels; otherwise one may not care for the results, and become exhausted anyway. This may be a complicated subject, unlike the present one which is perfectly simple; perhaps not so simple is the general influence of these ideas.

Contrary to what we read in textbooks, there is not and cannot be a representation of a social optimality in any usual market equilibrium model. This is obvious; however, though something like this has been said many times, it seems not to be acknowledged. Something else not so plain has a stronger influence. Bertrand Russell said "repetition is not a form of argument" but he was speaking maybe as a logician and thinking what we know, that repetition is a form of argument—a powerful one!

2.3 Pangloss

A choice has the form of a set with a single element picked out of it. One might question about the distinction of the element—what does it have that the others do not? That it has been chosen and the others have not is impressive: the other points seem to be losers. Then the point is *optimal*—in a sense, which makes the best of all possible worlds of Dr Pangloss, or the optimum of general equilibrium, or the paradox of the "Voting Paradox", or the revealed preference of the bundle of goods bought over all those that might have been bought instead with the same money.

'Optimum' is a term that has a great part in economics, so the sense of it is important. Ordinarily it signifies the best option for a specific purpose, by a criterion related to that purpose. There can be no reservations about that, and adherence to common usage should prevent any different meaning being given to the term, even in some special application. Where a choice is to be made, 'best' means 'chosen', since weighing alternatives as better or worse is only done in order to make a choice between them. There is a way of comparing alternatives, which exists separately in advance of the matter of making a choice and then comes to bear in the choice. Consider, for instance, wanting a heavy stone to serve as an anchor, the heavier the better, and looking around for the best stone, making comparisons. The stones had weight before that need for making a choice arose and regardless of it, and certainly before the optimal stone was found. A disturbing contrast is in 'optimality' cases of economics. An adjustment must be made, and here what is judged to be common usage will be adhered to. There might be an error in the judgement but at least the locus of it will be clear.

Acting so as to achieve the maximum of something has been offered as the definition of rationality. A first question that comes to mind concerns what is actually being said. Does it matter what is being made a maximum? If not then the function that is zero everywhere and thus also a maximum everywhere would serve well. If a strict maximum is wanted, so as to have a full explanation of the uniquely chosen object, a function that is one somewhere and zero elsewhere will be a strict maximum and *optimal* anywhere one wants. Such speculations cannot be part of the meaning of rationality, but still there is no guidance for knowing what is wanted.

Even if we put aside all the problems associated with choice and preference at the individual level, the transfer of the model for an individual to an arbitrary collection of individuals, found in welfare economics, should give us pause. Such a transfer expresses something like the *volontée générale* of the eighteenth century, associated with a collection of individuals being so settled together in some way that it amounted to a unified organism representing an individual of a new order, with a will encompassing all the individual wills. Now we have the same idea, but it involves an arbitrary collection, an abstract set, since nothing is spelt out about the members and their relationship to each other that produces the wonderful result. Modern theories claim to be explicit and to work with models in which everything that is used is always said in advance, if necessary by means of unambiguously stated axioms assisted by a free use of mathematical notations. They never pretended to do that in the rational eighteenth century—in a modern dress we have been taken back earlier!

2.4 Historical

Walk ever on the path of truth—with a sneer

Voltaire to d'Alembert

I have tried to understand what it is that Adam Smith's "invisible hand" is supposed to be maximizing

Paul Samuelson
"Maximum Principles in Analytical Economics"[32].

The idea of pursuit of the optimum, the sorting through of possibilities for some purpose to find the best, is understandable and commonplace. But along with it are doctrines about an optimum with a global reference produced without any intervention from ourselves. It is taught that a general economic optimum is associated with perfect competition. In another offering—with differences, though they appear not to matter—the optimum belongs to general economic equilibrium, or to a competitive equilibrium, though in this case the competitive seems to do no work and to be simply tacked onto the equilibrium, keeping up appearances in echo of the old doctrine where the competition seems to be important and is spelt out carefully at length. These matters are not in themselves understandable, but how such thinking ever came to be might be found out. That would be useful not only because of the classic cases, but also on account of fallout elsewhere.

A clue is found in historic simultaneity, and other coincidences, with the 'Optimism' of Leibniz. This was ridiculed by Voltaire and is now without influence as such, but it seems to have found a niche in economics where it has been able to survive with better protection. Leibniz, in his Théodicée (1710), propounded the doctrine that the actual world is the "best of all possible worlds" chosen by the Creator out of all the possible worlds which were present in his thoughts by the criterion of being the world in which the most good could be obtained at the cost of the least evil. This is the doctrine known as Optimism; in its time it drew a great deal of attention and is famous still. Voltaire's *Candide, ou l'Optimisme* (1759) with the well-known character of Dr Pangloss was "written to refute the system of optimism, which it has

---

[32] Nobel Memorial Lecture, Stockholm, 11 December 1970. In *Les Prix Nobel en 1970*. Amsterdam and New York: Elsevier. Reprinted in *Science*, 10 September, 1971.

done with brilliant success." All this and further information is in the *Oxford English Dictionary*. It was Leibniz who introduced 'optimum' as a technical term on the model of a maximum, and it first came into a dictionary in 1752. We are told:

> The optimism of Leibniz was based on the following trilemma:- If this world be not the best possible, God must either,
> 1. not have known how to make a better,
> 2. not have been able,
> 3. not have chosen.

> The first proposition contradicts his omniscience, the second his omnipotence, the third his benevolence.

The arguments about the economy are not quite like that. Instead there is a page of calculus, promising infinitesimal precision. It matters not about what, the results are the same. This is a parallel of the Maximum Doctrine that came into economics with Fran◊ois Quesnay and the Physiocrats and flourishes still. It is impressive to find Quesnay's Economic Principle "greatest satisfaction to be attained at the cost of the least labour-pain" perfectly represented in Leibniz's doctrine *vis-à-vis* the Creator's choice criterion. The senseless double optimization, found again with the "greatest happiness of the greatest number" formula, is avoided in the Pareto Optimum. This is not an optimum in the sense intended by Leibniz, even though he abused it, which continues to the present as the understood proper usage. But calling it an optimum shows respect for the old story. Under Pareto Optimism, with regard to the good and evil of the world, there would be the greatest good attainable with the given evil, and the least evil suffered for the good. Begging the main question by a cost-benefit analysis, suitable to mortals who have to get on with the job but no doubt contrary to the law of Heaven, Creation would have been delayed by the need to make a choice between points in the good-evil 'possibility-set'—as we would now say. Leibniz omitted a criterion for that. Were there a marginal price to resolve the matter, with the return of good for evil diminishing to a point of equilibrium, the economic analysis of Creation could have gone further with a use of the new Calculus. There could also have been discourse about the price, the author of it, and why it was not better, or worse.

2.5 Another report, and Pessimism

By another report, a virus landed on Earth in a meteor and the life that we know emerged through the effort to create a more hospitable environment. The important question then is whether our proper duty is being performed *optimally*. Neglect of the Virus Welfare Function only shows the ignorance that prevails about a fundamental matter.

More on the side of Pessimism, a worry brought forward recently, with a formidable display of erudition in scientific formulae, is Entropy. From Steam Engines, it went into Poetry—and now Economics. It is excellent for poetry, where there is no need for Boltzmann's equation. Now it comes into economics bolstered with all possible equations and a disturbing message: the entropy of the universe is increasing, everything is going downhill, bound to fall apart, final degradation is inevitable, and one is ignorant not to know it. This seems to be the 'Entropy Law' according to the recent innovation in terminology. It confirms the worst suspicions of some ecologists and others about reality, and gives cheer that truth is revealed at last to properly intimidated economists. There has been a stunned silence in the economics profession proper, but a few words by Harold Morowitz, a molecular biochemist of Yale University, serve well as a complete comment[33].

2.6 Important nonsense

The impossible 'happiness' formula is now known mostly as a Marxist slogan. But it had an early origin, as does the model for its illogic which came from Leibniz, entered economics with Quesnay, and was accidentally given a new though more subdued life by Pareto, which it still has. P. P. Wiener (1973) attributes the formula to Francis

---

[33] Review of Entropy, a New World View by Jeremy Rivkin and Ted Howland, in *Discover*, January 1981, 83-5.

Hutchinson (1694-1746), the teacher of Adam Smith. Its classic attribution is to the Utilitarians, and Marxists must have borrowed it from them. According to I. Philips:

> John Bowring says in his Deontology [1834, p.100] that Jeremy Bentham recalled how on a visit to Oxford in 1768 he had first come across the phrase "the greatest happiness of the greatest number", in Joseph Priestley's Essay on the first principles of Government, published in that year, 1768. It was from that pamphlet [Bentham said] that I drew the phrase, the words and import of which have been so widely diffused over the civilized world. At the sight of it, I cried out, as it were in an inward ecstasy like Archimedes on the discovery of the fundamental principle of hydrostatics, $Ευρηκα$.

Here is another thought, bright with the free market devotion:

> Edgeworth's pleasure machine assumption bore wonderful intellectual fruit it could be shown—with all the irrefutability of the differential calculus—that in a world of perfect competition each pleasure machine would achieve the highest amount of pleasure that could be meted out by society.

Robert Heilbroner
The Worldly Philosophers
(5th Edition, p. 172)

Enjoyment of the wonderful fruit should, in this case, be spoiled by a suspicion of worms. What is all the irrefutability of the differential calculus? Is it like irresistible authority of the Chain Rule? Or final truth in the Infinitesimal, unphased by digital diversions? Or the incomprehension and boredom of all those readers who give a passing glance at the exhibition of machinery and then get on with the text?

We should do that first, since the outer skin of this fruit is not without blemishes. We are faced once more with the Leibnizian nonsense, expanded into *n* dimensions. That ought to be a relief, since now there should really be no need to go back to the skipped-over calculus after all. However, belief there is relief is feeble *optimism*, a dream of *rationality*. For the particular calculus turns up in countless textbooks—at least we should know now where it started.

2.7 Welfare again

> If we continue the traditional identification of rationality with a maximization of some sort, then the problem of achieving a social maximum derived from individual desires is precisely the problem which has been central to the field of welfare economics

Kenneth J. Arrow
Social Choice and Individual Values, 1951

This statement has influenced a generation or two, so even if positions have changed in the meanwhile it deserves a comment. For some, possibly everyone, the 'traditional identification' starts here. In any case 'rational' has diverse uses not all to be killed off in the one stroke. Arrow's own use is connected perhaps with another and through carelessness might be taken to be the same. That has to do with the doctrine of free will where man, being endowed with reason, has to choose between good and evil. Man knows good from evil but the choice is still a problem. In welfare economics it is rather the other way round: the determination to choose the best, or maximum, is fully taken for granted; the problem, instead, is knowing the better from the worse. A fair connection might be found if the choice between good and evil were as simple as optimization, but apparently it is not, and dispute is possible. Dr Pangloss was hanged (instead of being burnt—because it was raining!) for speaking about the matter. And poor Candide was beaten just for listening.

The brevity of the above passage conceals a complexity of which this matter of use of a word is only a part. A significance of bringing in rationality at all has to be known. Anything linked with rationality is usually rated a good thing, though the importance of it can be exaggerated. In any case, what is brought before us is something social—never mind what—"derived from individual desires". A sense that can be made out is that the derivation is in some way democratic, with the result for society being decided by its individual members—for instance by taking a vote, though nothing so commonplace is contemplated. One could hold on to this idea as a possibly clear element in the matter. Rescued—or even not—from the quagmire made by company with rationality, maximality, welfare, and so forth, it has helped stimulate the attention given to democratic decision processes.

But we should revisit the quagmire. One hears about the 'group mind', though it is difficult to be rational about it, and in any case no one ever said it was rational. The group mind syndrome is manifested in this very subject, and that is how the irrational phenomena in it ought to be understood.

In the 'traditional' adherence, rationality is associated with mind or thought belonging to individuals. However, after maximization has been blessed with the name of rationality by the rhetorical "if then " we find it promptly applied to the group, any group. We had that already in the beginning with the antique Maximum Doctrine of the Physiocrats, and then with modern welfare economics. Now we should have it still, but with a better modern, and at the same time properly traditional, conscience, giving complete courage for what follows. That contains mathematics which is unusual and original in itself, so as to give interest regardless of what otherwise it should be about. An accidental effect is to enhance the credibility of ideas offered at the start.

The voting paradox has prominence, but it is a paradox only if one sees the elected candidate—surely "derived from individual desires", or votes at least—as not simply elected but also best, a social maximum. Since the paradox is not made into a lesson for not seeing elected candidates that way, it becomes the opposite and reinforces the simplistic optimization way of thinking which is important for welfare economics.

A giver of solutions to problems, the mathematical mode is also a problem itself, because of the scientific aura. No strategy is suggested here, but something parallel involving the same psychology has been well expressed by Harold Morowitz (1981): "A popular strategy in modern salesmanship is to associate an impressive scientific term with a product. Thus 'protein' has been put into shampoo, 'nucleic acid' into hair rinse—and 'entropy' into economics and sociology."

A group, as understood in choice theory, should be a model that involves individuals and their connections, not just an abstract set. The model should be explicit about its features, so that it is known what is being dealt with: there are the individuals and, moreover, there is what they have to do together. Here the matter is just terminology, but there can be obscurity in arguments dealing with a group about what it is that makes the individuals into a group.

In the familiar economic model, there are individual agents whose connection with each other rests solely on the fact that they trade goods at certain prices. They take notice only of prices and encounter each other only because, so we understand, wherever there is a buyer there must be a seller and conversely. These individuals, though a group by virtue of the transaction connections, have no purpose or other government other than their own separate ones, by which they voluntarily enter into the transactions; the only interface between them is the price. In the model they have no community but prices—no political connection, and no other expression of a common interest. The model even lacks the terms that might provide a definition of group welfare and give it a significant function. But still group welfare is talked about. One should wonder how this is possible. It might be possible to envision a model that included some concept of group welfare, but it would be a different model.

In microeconomics, an economy is a model consisting of a group of individuals who form a system through their transaction relationships. Political theory might take a political body to comprise a group of individuals bound together by a constitution. For purposes of ideal discussion, economic and political aspects may be isolated from each other, even though in experience they are bound together; desiccated idealizations are better for purposes of abstract discussion. We have theory that deals with the characteristics of groups of individuals making group decisions based on individual decisions, as in democratic processes. The theory might relate primarily to politics, but it has come to be applied as well to economics, where, though there can be doubts that it should, it is linked with welfare theory.

Groups of various kinds are found in the world of experience, and they make group decisions—a school of dolphins, flight of geese, sport team, military unit, biological population, society of cells in an organism, and so forth. One can compare a political body or economic system with such groups for similarities and contrasts. Some decisions taken by these groups may be comprehensible and others mysterious, but in any case we would not think of interpreting all their decisions as simple optimization.

3 Keynes on "National self-sufficiency"

3.1 Free trade

But when we wonder what to put in its place, we are extremely perplexed.

Keynes

International relations have been well known for instability, resulting in breakdowns that lead to military confrontation. Today counterbalancing these traditional tensions are additional ecological, environmental, and population concerns, as well as the intense involvement in trade. There may be hope for greater stability, but no guarantee.

There is the familiar pattern where nationalities, markets and religions bind masses of people together—and contribute to conflicts. States require a basis of security for viability. Hence outside threats are matched by measures of defence, which in turn give a capability for offence, feeding new needs for defence, creating an expanding, exhausting cycle. This is the typical destabilization pattern, and economic relationships can fall victim to it when things go wrong.

The expanding spiral has been at the root of difficulty about disarmament, making it not quite feasible. But the major powers now have the deterrence of the bomb, and they suffer exhaustion from the demands of the spiral without having reached the point of war. There is something like the appearance of a new situation.

It may seem that almost anything can be thought or said in economics and probably has been. Now there is the suspicion that, as a result of volatilities enhanced by electronics, which allow massive movements of capital almost instantaneously, perhaps just by chance, an instability of a new order may be creeping into the system—one that may at some point test the robustness of the system, or demonstrate its fragility.

Trade seems to be the order of the day. The current emphasis is on the gains from trade, expansion is an understood condition for prosperity, and free trade (nebulously, whatever it should mean, since in practice trade is always hedged in by all sorts of rules and regulations) is held up by the more determined traders as the ideal.

Few try to imagine what a world of truly open markets and free trade would be like—the befuddlement with global noise, the unrelieved stress from adjusting to movements and dealing with clever manoeuvres. Alternatively, there is the view of the market as invisible hand or as a medium for coordinating supply and demand. According to this view, the market is just a machine that does not notice what it does. It is driven by gain and not everything that happens is open and friendly. Hence understandably, when it is possible, wherever there is a sovereign community and not just a trading post, there is resort to 'protectionism'.

It is possible to be in favour of peace without being sure what should be done with it. George Ignatieff[34] argued that cutting back armaments can serve a country's competitiveness in the world market. If that should be the result of peace, it could be a matter of jumping from the frying pan into the fire. The problem of peace, which once meant how to attain peace in the first place, now may take on an additional meaning. Peace once attained will not be without its own difficulties.

The building of self-sufficiency in some sense, or viability without completely haphazard dependence on others, is an alternative worthy of consideration by those who would know how to enjoy it. It could affect development policy, which so often promotes trade interests under the guise of aid.

Protecting borders is called defence, something understandable and commonplace, even good. However, protecting an economic

---

[34] Holtom Lecture, Ottawa, 6 February 1989.

community is 'protectionism'—not so good! Yet the alternative to
economic protectionism is to be exposed to arbitrary happenings that do
not serve home interests, that absorb attention and require continual and
costly adjustments beyond the capacity of many, who then fall into a
disadvantaged position. The promotion of free trade as an ideal and an
unqualified good, besides the falsity in it, in effect covers over issues
that should require deliberation. Such teachings in any case tend to be
disregarded by those with practical responsibilities, so it is to some
extent inconsequential. The doctrine of free trade is like those
portentous and hazy formulae or slogans that bolstered claims of the
recently abandoned communist ideology. Nonetheless, as in that case, it
does have significant influence and invites contrary thinking. For that
the 1933 article of Keynes does service, even though he had been
concerned about peace at a time when its preservation seemed unlikely.
He reopens ideas where teachings aided by objectionable argument
would have them closed. Unless indicated otherwise all following
quotations are from that article.

3.2 External affairs

> I thought England's free-trade convictions, maintained for nearly a
> hundred years, to be both the explanation before man and the justification
> before heaven of her economic supremacy.

Keynes

While once one would usually have gone to market for a few things
largely of local origin, today we are in the midst of a global system, one
without respect for locality. There is nothing wrong with markets as
such, where goods are traded or bear a price. They are not the discovery
of modern economics but have always existed, coming into being
wherever there has been some protection of property. Traders, as such,
must find the market a good thing, but no one long ago said it was
'optimal', whatever that may mean. Access to a market opens
opportunity. There may then be a temptation, but whether or not to
yield to it should still be a question.

According to Keynes, "to shuffle out of the mental habits of the nineteenth century world is likely to be a long business". These habits still persist and one may wonder whether the shuffle can have a quicker pace. They were the habits of the undisputed economic rulers of that era, Britain and then the United States, that gave confident approval to the "survival of the economically fittest", themselves[35]. In view of changes in the world, some other thoughts should probably be admitted.

> I sympathize…with those who would minimize, rather than with those who would maximize, economic entanglement between nations.
>
> **Keynes**

This idea is at odds with normal thinking, and certainly with surrender to indiscriminate globalization. But the guard at a frontier does not only watch for an invasion of troops; anything that would cross can be inspected. When the division of labour is freely extended internationally, the frontier is opened and an element of sovereignty and security has been given up.

Communities are formed by people living and acting together, with common interests for pursuit and protect. They share in the care and defence of a territory, and enjoy the resulting security, create a market for the division of labour, or are united in other ways. However that may be, a community, perhaps just a village, even a club, is marked by separateness from the outside.

In the dependence of welfare of a country on external relationships, defence is a branch, and there one thinks first of the military. This takes away from other aspects, such as economic. However, in the age of deterrence through mutually assured destruction, military exercise is restricted, if not prevented, and relationships between countries are confined more to economics. Functions of defence are transferred to the economic sphere, and inert military weight can become a debility.

---

[35] The Japanese are of course successful traders, even a model nowadays, though they seem not to be at all doctrinaire about it.

In his 1933 article Keynes is concerned about defence where the military is not given high priority, and about peace. This is not 'Keynesian economics'; rather it seems to be an isolated offering that he did not pursue further. Most of what has happened more recently is in the opposite direction, but that does not alter the interest of it.

The self-sufficiency envisaged cannot, of course, be understood in any completely simple sense, but it includes the idea that it is better not to have a way of life where one is at the mercy of others for important essentials; for these, Keynes give importance to proximity. Priorities may produce stages of community, anything outside at any stage being more expendable than anything within. Since the Earth itself is not expendable, there has to be some pulling together at that, the final stage.

3.3 Proximities

Let goods be homespun.

Keynes

The Earth is a community now that any part of it can be reached from any other in a few hours, and it is especially that since its destruction would affect everyone. People become associated by proximity; friends, and enemies, are usually neighbours. There are many kinds of distances, but geographical location reduces them, making it, as it has always been, especially important. Territories have been delineated by oceans, rivers, and mountains, beside other accidents of history.

A definition of self-sufficiency cannot be simple at this particular time, though new factors might now serve the idea better. Images of the old city-state, or the Virgilian agricultural estate, may not fit entirely though they have something to offer. But in any case it would be helpful now to understand Keynes's concept of national self-sufficiency, and why he advocated it. In particular, Keynes proposed proximity of producers and consumers, and of the owners and operators of productive facilities. This is at variance with wisdom of 1933 and today.

Proximity may be understood to serve sovereignty, which usually, in the first place, has a territory as reference, and the security of some degree of sufficiency. Regional deficiencies can be diminished by trading in a community; but a community as a whole may itself have deficiencies. Hence communities have reasons for getting together, as it were in a hierarchy providing progressive extensions of home. Collective concern for the Earth, the final home extension, imposes a community over all others; whatever the deficiencies there, they have to be lived with, since there is (as yet) nowhere else to go.

Keynes remarked on various political experiments going on around the world. He preferred self-determination over mutual interference, whereby communities might sink or swim as they choose. The needs of community may produce some convergence of elements, but the divergence into variety is in the order of things and has its claim: as might be allowed, every specific locus is unique and has its peculiar entitlement and sovereignty.

3.4 Distances

Remoteness between ownership and operation is an evil.

Keynes

Dependence on others in some vital matter, especially when it is not reciprocal, leaves one disadvantaged and exposed. The destruction of the Adelaide market (see last section), which can be interpreted in other ways, is a good example. The oil embargo demonstrates the case of import-dependence and its hazard. Similarly there is export-dependence, at an unfortunate extreme in the case of single-crop economies. Economic vulnerability can lead to unsettlement and wreckage comparable to that from military assault. Yet economic defence is not taken as seriously as military defence, and not given the same approval.

Keynes questions the "great concentration of national effort on the capture of foreign trade ... the penetration of a country's economic structure by the resources and the influence of foreign capitalists" and the "close dependence of our own economic life on the fluctuating economic policies of foreign countries." He finds nothing here to serve stability and peace. In a scheme of things "which aims at the maximum international specialization and at the maximum geographical diffusion of capital wherever the seat of ownership", the "protection of a country's foreign interests, the capture of new markets" and "the progress of economic imperialism" are unavoidable. Coherent proximities are spoilt, and life is exposed to remote disturbances and dreadful complications. He goes further, objecting to other incoherencies and distances:

> The divorce between ownership and the real responsibility of management is serious... when, as a result of joint-stock enterprise, ownership is broken up between innumerable individuals who buy their interest today and sell it tomorrow and lack altogether both knowledge and responsibility towards what they momentarily own... I am irresponsible towards what I own and those who operate what I own are irresponsible towards me.

He illustrates this idea with "the part ownership of A.E.G. of Germany by a speculator in Chicago, or of municipal improvements of Rio de Janeiro by an English spinster." In allowing that "There may be some financial calculation which shows it to be advantageous that my savings should be invested in whatever quarter of the habitable globe shows the greatest marginal efficiency of capital or the highest rate of interest," he brings forward neglected factors "which will bring to nought the financial calculation." He cedes possible merit, in its time, to the general presumptions regarding the fundamental characteristics of economic society that prevailed since the nineteenth century, and declares, "I become doubtful whether the economic cost of national self-sufficiency is great enough to outweigh the other advantages of gradually bringing the producer and the consumer within the ambit of the same national, economic and financial organization." Then he further states:

> Experience accumulates to prove that most modern mass-production processes can be performed in most countries and

climates with almost equal efficiency. Moreover, as wealth increases, both primary and manufactured products play a smaller relative part in the economy compared with houses, personal services and local amenities which are not subject to the international exchange; with the result that a moderate increase in the real cost of the former consequent on greater national self-sufficiency may cease to be of serious consequence when weighed in the balance against advantages of a different kind. National self-sufficiency, in short, though it costs something, may be becoming a luxury which we can afford if we happen to want it.

If that was at all true in his time, it could be more so now.

Ecology brings forward the notion of the relationship between people and their territory, fostered by decentralization towards small units. People are not so much owners as caretakers. Resources are limited, so human needs must be limited too; insatiability comes from an increasing population and competitive pressures. In the absence of such raw factors that undermine it, and given the needed character for the undertaking, any people can, in their own way, build from subsistence towards satiation—a state that meets their own needs while not undermining those of the Earth.

3.5 High points

The intelligence which proceeds not by hoping for the best (a method only valuable in desperate situations), but by estimating what the facts are, and thus obtaining a clearer vision of what to expect.

Pericles

Not only capitalism, with its global market free of any regulation but its own, but also Marxism has its high point in the "withering away of the state", under strange assumptions about the nature of man, discussed by

J. Alvey.[36] The two share a similarity on this point, even though the one withering may not be quite the other. As for any principle that may be drawn from Keynes's own article, if there is one, it is proximity, it is having the sense needed for coming down to Earth—an important proximity. He gives this principle several applications. It may not join with the sublimities of Leibniz and Hegel, but in sober times it may be enough. Keynes seems to have in mind an order different from what we have, even though he does not provide details.

What is the purpose of Keynes's order and how should it come about? He submits it is not an end in itself but is "directed to the creation of an environment in which other ideals can be safely and conveniently pursued." He sees the emergence as gradual: "It should not be a matter of tearing up roots but of slowly training a plant to grow in a different direction." Considering the field for the training, this may be an optimistic way of putting it. With regards to a future order, time with its normal dispensation of disaster may bring it about in the shape of repairable fabrics that survive. But Keynes seems to think that, after "estimating what the facts are", there is a path that should be taken anyway.

Clearly his views could have no impact at the time. He seems to have disregarded for the momentum of the present, choosing instead to give his view of what economics must eventually be about. His ideas relate not just to the concerns he had in 1933, but to the present day and the future. "Estimating what the facts are," for Keynes, most immediately concerned the oncoming world war, while now the world-encircling realities have to do with interdependent factors of ecology, environment, and population.

3.6 Interests

Current issues were highlighted in a debate on the public television programme "American Interest", one side taken by the chairman of

---

[36] "The relevance of Marx's assumptions on the nature of man for his economics." Research Paper No. 314 (April 1987), School of Economic and Financial Studies, Macquarie University, North Ryde, N.S.W.

Citizens against Foreign Control of America (C.A.F.C.A.), June-Collier Mason. The opponent advanced the usual unreconstructed dogmas about gains from trade, while Mason held that not everything was for sale, or could be bought. It seemed a questionable evolution when states vie with each other for foreign investments or purchases, even though one knows the usual arguments: to create jobs, stimulate the economy, transfer technology, and so on. There could be a need met and a gain obtained, but one can wonder where this will lead if the drive continues—as with drug dependence. There must be things that an independent people should do for themselves, and the spectacle makes any concept of self-sufficiency seem empty. But we may just be witnessing the *reductio ad absurdum* of the unqualified free trade indoctrination, preachers themselves now being principal victims.

Next to this phenomenon, for contrast, one can contemplate how Japan, now the trader, once withdrew from the international scene for about 260 years. It took gunboats to make them open up—for trade, of course.

The old forms of national power, backed by the military, have not been attenuated but transmuted and channelled elsewhere, preserving the usual potential for coercion and conflict. Economics is as serious as war and closely related to it. The Peloponnesian War was precipitated by the denial of the Athens market to an ally of Sparta, and it took an American gunboat to persuade the Japanese into trade relations. Apparently Keynes was not satisfied with the concept of what constitutes peace and was looking further.

His exploration may be understood better today than in 1933, for many reasons. It is not related to "Keynesian economics" which Axel Leijonhufvud[37] has distinguished from "the economics of Keynes", or to anything that one usually associates with Keynes; rather it has to do with a neglected item.

---

[37] *Keynesian Economics and the Economics of Keynes*. New York: Oxford University Press, 1968

4 In Adelaide (or anywhere)

Not only do I have an apparently eccentric fondness for an ancient economic institution, but I am led to strange political conclusions.

Michael Symons
"In Adelaide, the collapse of the free market"
Sydney, NSW: *Times on Sunday*, 3 May 1987

An important kind of market is the local one, associated with and requiring the protection of a specific community. Giants stride the global market seeking entry where they choose, destroying local fabric wherever they go. Their unhampered access can only be explained by the unguardedness of victims.

Michael Symons (1987) describes an illustrative case where the East End market in Adelaide, South Australia, was being demolished and moved to the outskirts. The change would serve the interests of large-scale producers, merchants, supermarkets, food processors, and the developers of the valuable city site. At the same time, "food [would] lose cheapness, freshness, quality and seasonality." After some pressure, the Edwardian facades were retained, as a "heritage case". However, "a market is more than a building, being a key gastronomic, agricultural, economic and civic institution". Adelaide is to be locked another step into the national and international distribution system:

> Just as the shift will hasten the demise of small market gardeners and orchardists, so too will more greengrocers be forced out of business by the greater distances, changed hours and much higher rents. Supermarkets, which presently by-pass the East End market, can be accommodated at the new 32-hectare centre.
>
> So, we will lose a few more primary producers, market workers and corner greengrocers: Adelaide citizens will get more expensive, older and lower-quality fruit and vegetables (try shopping at our supermarkets now) and, more profoundly, we will further lose seasonal and regional variation in our food.

The costs and benefits and their incidence are well outlined in this description. The global market acts as a destructive solvent, corroding and carrying away local particulars. Its work could have been stopped, but the accepted teaching has given it an inevitability and legitimacy, numbing thought and persuading submission. How else can one explain "why farmers, who so often complain they are ignored, aren't protesting at the loss of their city presence, and their livelihoods? Why aren't small retailers rioting against unfair competition, and unemployment? Why don't governments care about eaters?"

The destruction of *Les Halles* markets in Paris is compared with razing Notre Dame Cathedral, leaving Parisians with more expensive, inferior produce: "A vital cuisine is derived from the basic level of the myriad activities of individual operators—not from food giants. So when the East End market goes the way of *Les Halles* and Covent Garden, Adelaide will lose a colourful centre and, more importantly, the original free market."

As said by Fernand Braudel, "markets are the *raison d'être* of towns. They are the birthplace of our economy." A society fails when it allows its markets to be destroyed because of excessive love of polemicist theoreticians. As with the monotheistic dedication these envisage a single universal being, the Market (these days global though hitherto not so specific) as ground for all economy and endow it with the divine Optimum. It was supposed the age of faith had passed, but apparently not.

Zealous devotees of the market system, missing the simple point of it and pursuing a new theory dressed with equations and permeated with "unjustified scientism" and worse, may be its enemies. Thinking of a market which is "perfect" and "optimal" (for whom and in what respect?) they advocate a chaotic abandonment of restraint and offer less in the way of discipline that could serve the system and its genuine welfare aspects.

6
Conclusion

The conclusion starts with a reminder of how we begin:

> IT IS VERY PUZZLING that the most important
> problem in the world is without attention,
> hardly talked about, just to mention it would be
> "not politically correct".
> It was without mention in the recent Copenhagen
> Conference.
> Is this a symptom of the essential human final defeat?
> Is not the general commotion about climate change
>   rather a distraction?
> This book gives a new opening for the subject.

The Copenhagen Conference came after publication of the book and helped prompt this declaration. Now following an unavoidable prolonged delay, in the moment we are at last about to take up about Conclusion, we read in large letters[38]

Global population study launched by Royal Society

The UK's Royal Society is launching a major study into human population growth and how it may affect social and economic development in coming decades

Occasion for surprise and a pause ... !

---

[38] http://news.bbc.co.uk/2/hi/science_and_environment/10578484.stm

Then for another pause, "What is the Nature of the Population Problem and How Can It Be Solved?" by Amartya Sen [39].

Our own primitive ideas do not require attention to new studies, or to history since Malthus. We do not have to read the highly readable and well informed essay of Sen or await findings of the Royal Society study.

The foundation for our approach, as should be for any, is 'Fundamental Preparation' FP, made by awareness of everyone about the population problem, FPA, and availablility of family planning facilities, FPF, worldwide to reach the entire population everywhere. There should be no delay proceeding unobstructed by pervasive hangover from prehistoric culture.

The 'Raplacement Principle' RP requires the reproductive facility for all humans to serve only for replacement. This is central to the approach, where FP opens the way to follow RP in practice and for good reason.

Now we take choosers of RP to make 'the World Family' WF. There is distance in this concept from population controle or government, instead proximity to free choice.

For the two parts of FP, FPA and FPF, while the second has substantial ordinay experience related to it, the first is less straightforward. A brief document is wanted, say one page, for easy copying and distribution, and effective for its purpose with the generality of humans. At any time there could be several versions of such a document in use. Real talent is required to produce one. An appeal should be made to Amartya Sen to offer one, An auxiliary document could act as an invitation to join WF.

While the matter of climate change makes an essentially very difficult subject with a scarcity of certainties, population growth is rather the opposite. Here there is the great certainty that if any part of the world were to join WF, population growth there would stop absolutely. In other words humans have control of population but not of climate.

---

[39] Harvard University, delivered at Keio University, 1995

Effort should be concentrated more where it is known there can be effect.

This raw WF concept touched so far for its introduction can nonetheless have wrinkles. A case to consider is where replacement is given an enlarged meaning where procreation rights are voluntarily transferable.

For instance one family may for purely personal reasons be pleased to give up the right for producing one child to have it accepted by another that may even already have two. In any case transfers can take place purely by private agreement between those involved.

For another case, the one family may require compensation to be provided by the other, the one poor and the other rich, so here is some facility for dealing with poverty. For families that live on handouts it is suitable that WF registration be required.

The poor have large families and also would have interest to join WF for the material benefit available from accepting a small family size. Hence the greatest population reduction will be derived from poor families so **povery elimination and population reduction go hand in hand**

A bank for WF, the 'World Family Bank' WFB, could collect penalties from families that overstep their rights and provide compensation to others doing the opposite. It could also serve families interested in finding a transfer in harmony with special conditions

With the so-called Population Problem there is usually some presumed reference consisting of a concentration of humanity located somewhere, where there is a Government, with a Population Policy. and the Royal Society study could surely further Government with that old-fashioned factor in its grim history.

With our imaginary experiment we have no such things. There is just one thing, which is to have humanity in the large and everywhere fall in with a particular notion of Social Correctness (as it were Voluntarily without resistance to that all-powerful influence).

Bibliography

Afriat, S. N. (1965). People and Population. *World Politics* 17 (3), April, 431-9. *International Political Science Abstracts UNESCO.* Japanese translation with foreword by Edwin O. Reischauer (US Ambassador to Japan): *Japan-America Forum* 11, 10 (October 1965), 1-28.

(1980). *Demand Functions and the Slutsky Matrix.* Princeton University Press, Princeton Studies in Mathematical Economics, 7.

(1987a). Economic Optimism. Economics Department, Stanford University, 21 April; and in Australia, May-June: ANU, Macquarie, Melbourne, Sydney, Newcastle.

(1987b). *Logic of Choice and Economic Theory.* Oxford: Clarendon Press.

(1988). On Trade, and Self-sufficiency. Institute of Economics and Management Science, Kobe University, 22 April 1988. Revised versions: *Research Paper* No. 326 (April 1989), School of Economic and Financial Studies, Macquarie University, New South Wales; *Quaderno* No. 284 (March 2000), Department of Political Economy, University of Siena, http://www.econ-pol.unisi.it/quad2000.html; included in Afriat (2005).

(1988). Optimism from Leibniz to Modern Economics. Sophia University, Tokyo, 27 April.

(1994). Market & Myth. 2nd International Conference, The Society for Social Choice and Welfare, University of Rochester, New York, July 8-11.

(1999). In the Economic Context: Concerning Efficiency. Symposium 'On Effectiveness', Centre for Interdisciplinary Research on Social Stress, San Marino (Republic of San Marino), 20-25 May 1999. *Quaderno* No. 254 (May), Department of Political Economy, University of Siena.

(1999). Market Equilibrium and Stability. *Quaderno* No. 264 (September), Department of Political Economy, University of Siena.

(2005). *The Market: Equilibrium, Stability, Mythology.* Foreword by Michael Allingham. Routledge.

.All Party Hearings Report (2007). Return of the Population Growth Factor: its impact upon the Millennium Development Goals. *All Party Parliamentary Group*, January.

Allaby, Michael and Peter Bunyard (1980). *The Politics of Self-Sufficiency.* Oxford University Press.

Alvey, J. (1984). John Maynard Keynes and his relevance for today. *Economic Analysis and Policy* 14, 1 (March), 98-118.

(1986) Keynes: fifty years later. *Australian Society*, December 4 - 5.

(1987a). John Locke's Theory of Property. *Working Paper in Economics* No. 62 (March), Department of Economics, University of Queensland.

(1987b). The relevance of Marx's assumptions on the nature of man for his economics. *Research Paper* No. 314 (April) School of Economic and Financial Studies, Macquarie University, Sydney, North Ryde, N.S.W.

Arrow, K. J. (1951). *Social Choice and Individual Values.* New York: John Wiley. 2nd edition 1963.

Beckerman, Wilfred (1974) *In Defence of Economic Growth.* London: Jonathan Cape.

Bell, Daniel and Irving Kristol (1982). *The Crisis in Economic Theory.* New York: Basic Books.

Boulding, Kenneth E. (1963). The Legitimation of the Market (mimeo). Lecture delivered at Rice University, Houston, Texas.

Bourne, Joel K. Jr (2009). The End of Plenty, *National Geographic Magazine* (June), http://ngm.nationalgeographic.com/2009/06/cheap-food/bourne-text.

Braudel, Fernand (1981). *Civilisation and Capitalism.* New York: Harper and Rowe.

Brecher, Irving and Donald J. Savoie (Eds.) (1993). *Equity and Efficiency in Economic Development: Symposium in honour of Benjamin Higgins.* Montreal: McGill-Queen's University Press.

Brown, Lester R. (1974). *In The Human Interest: A Strategy to Stabilize World Population.* New York: W. W. Norton & Co.

Cassidy, John (1996). The Decline of Economics. *New Yorker*, December 2, 50-60.

Conolly, Matthew (2008). *Fatal Misconception: The Struggle to Control World Population.* Harvard. (Review by Dominic Lawson, *The Sunday Times*, 18 May 2008,)

Dasgupta, Partha (2005). Regarding Optimum Population. *Journal of Political Philosophy* 13 (4), 414-42. www.econ.cam.ac.uk/faculty/dasgupta/pub07

De Villiers, Henri (1989). Quoted in *TIME*, March 27, p. 42.

Deane, Phyllis (1978). *The Evolution of Economic Ideas.* Cambridge University Press.

Ehrlich, Paul R. (1968). *The Population Bomb.* NY: Ballantine Books

Falk, Richard A. (1963). *Law, Morality and War in the Contemporary World.* New York and London: Praeger. Princeton Studies in World Politics No. 5.

 (1971). *This Endangered Planet : Prospect and Proposals for Human Survival.* New York: Random House.

 (1992). *Explorations at the Edge of Time: The Prospects for World Order.* United Nations University Press, Tokyo, and Temple University Press, Philadelphia.

Freedman, Ronald (ed.) (1964). *Population: The Vital Revolution.* New York: Doubleday.

Galbraith, J. K. (1973). *Economics and the Public Purpose.* Boston: Houghton Mifflin.

Gao, Jinbao (1987). The reform of the Chinese economic system (mimeo). Lecture delivered at Dunmore Lang College, Macquarie University, NSW, May.

George, Susan (1988) *A Fate Worse Than Debt.* Penguin.

Goldsmith, Edward (ed.) (1971). *Can Britain Survive?* London: Sphere Books.

 (1972). With Robert Allen, Michael Allaby, John Duvall and Sam Lawrence *A Blueprint for Survival.* London: Tom Stacey.

(1980). Thermodynamics or Ecodynamics. *Ecologist*, 178-195.

(1984). *The Social and Economic Effects of Large Dams*. San Francisco: Sierra Club Books.

(1992). *The Way: An Ecological World View*. London: Ryder.

Hansen, Bent and Girgis A. Marzouk (1965). *Development and Economic Policy in the UAR (Egypt)*. Amsterdam: North-Holland.

Hauser, Philip M., ed., *The Population Dilemma* (Englewood Cliffs, NJ 1963)

Hayek, F. A. (1974). The Pretence of Knowledge. Nobel Memorial Lecture, Stockholm, 11 December. In Hayek (1978), Chapter 2.

(1976) The Atavism of Social Justice. The 9th R. C. Mills Memorial Lecture, University of Sydney, 6 October. In Hayek (1978), Chapter 5.

(1978). *New Studies in Philosophy, Politics, Economics and the History of Ideas*. University of Chicago Press.

Heilbroner, Robert (1972). *The Worldly Philosophers*. New York: Simon and Schuster.

Higgins, Benjamin (1989). Equity and Efficiency in Development: basic concepts. In Brecher and Savoie (1992), Chapter 1.

Holdren, John P. (1990). Energy in Transition. *Scientific American* 263, 3 (September), 156-63.   http://www.econ-pol.unisi.it/quad2000.html.

Hutchison,T.W.(1977). *Knowledge and Ignorance in Economics*. University of Chicago Press.

IAAH (International Alliance Against Hunger) (2007). Resource Mobilization Strategy. *IAAH Secretariat*, September.

Ipsen, Eric (1997). Capitalism's King Spies Evil in Market-Mad Realm. *International Herald Tribune*, January 16, pp 1 & 6.

Johnson, Boris (2007). Global over-population is the real issue. *Boris Johnson's Office.*

Keynes, J. M. (1933). National Self-Sufficiency. *The New Statesman and Nation*, 8 and 15 July; *The Yale Review*, Summer; *Collected Writings* edited by D. Moggridge. London: Macmillan (1982), Vol. XXI, 233-246.

(1982). *The Collected Writings of John Maynard Keynes*, edited by D. Moggridge. London: Macmillan.

Kuttner, Robert (1985). The Poverty of Economics: a report on a discipline riven with epistemological doubt on the one hand and rigid formalism on the other. *Atlantic Monthly*, February.

Kuznets, Simon (1973). *Population, Capital, and Growth*. New York: W.W. Norton & Co.

Leijonhufvud, Axel (1968). *Keynesian Economics and the Economics of Keynes*. New York: Oxford University Press.

(1973). Life among the Econ. *Western Economic Journal* 11, 3 (September); reprinted in Dimand (1986).

Ligota, Christopher (1989). Stasinos quotation, Private Communication, Warburg Institute

Lovelock, James (2006). The Revenge of Gaia: Earth's Climate Crisis and the Fate of Humanity. New York: Basic Books.

(2009a). Private Communication, e-mail 22 April.

(2009b). *The Vanishing Face of Gaia: A Final Warning*. New York: Basic Books.

Lucas, F. L. (1961). *The Greatest Problem: and Other Essays*. New York: Macmillan.

Malthus, Thomas Robert (1798). *Essay on the Principle of Population*.

Marzouk, Girgis A. (1972). *Economic Development and Policies: Case Study of Thailand*. Foreword by Jan Tinbergen. Rotterdam University Press.

Mason, J.-C. (1987). The trade war in on; we must fight to win. *USA Today*, 1 April.

(1988). Foreign money is bad for USA. *USA Today*, August 17.

McKie, Robin and Caroline Davies (2008). Special Report, *The Observer* International Edition, Sunday 7 September.

McMurtry, John (1988). The Unspeakable: Understanding the System of Fallacy in the Media. *Informal Logic* X.3 (Fall), 133-50.

(1990). Education for Sale. *CAUT Bulletin* 37, 7 (September), pp. 11, 15.

Mishan, E.J. (1981) *Economic Efficiency and Social Welfare*. London: Allen & Unwin

(1986) *Economic Myths and the Mythology of Economics*. Brighton, Sussex: Wheatsheaf Books.

Morowitz, Harold (1981). Entropy, a New World View by Jeremy Rivkin and Ted Howland. *Discover*, January, 83-5.

Morowitz, Harold (1981). Review of Entropy, a New World View by Jeremy Rivkin and Ted Howland, in. *Discover*, January, 83-5.

Mudd, Stuart, ed., *The Population Crisis and the Use of World Resources* (Bloomington, Ind. 1964);

Nell, Guinevere (2008). Private communication, July.

Nicholson-Lord, David (2007). Citizens arrest: Tackling climate change is now a crusade, so what's stopping the simplest solution. *The Guardian Weekly*, 24 August, p. 40.

Olson, Mancur and Hans H. Landsberg (eds.) (1973). *The No-Growth Society*. New York: W. W. Norton & Co.

Organski, Katherine and A. F. K. (1961). *Population and World Power*. New York: Alfred A. Knopf.

Polanyi, Karl (1944). *The Great Transformation: the political and economic origins of our time*. New York: Rinehart & Co. Inc. (and Boston: Beacon Press, 1957).

Population Institute (1990). No matter what your cause—*it's a lost cause*—if we don't come to grips with overpopulation. *The Population Institute*, Washington DC.

Samuelson, Paul A. (1970). Maximum Principles in Analytical Economics. Nobel Memorial Lecture, Stockholm, 11 December. In *Les Prix Nobel en 1970*. Amsterdam and New York: Elsevier; reprinted in *Science*, 10 September, 1971.

Schumpeter, Joseph A. (1954). *History of Economic Analysis*. New York: Oxford University Press.

Scientific American (1990). Special Issue: Energy for Planet Earth. Volume 263, 3 (September).

Settle, Thomas (1976). *In Search of a Third Way.* Toronto: McLelland & Stewart.

Shimm, Melvin G. and Robinson O. Everett (eds.) (1961). *Population Control: The Imminent World Crisis.* Duke University Law School, Law and Contemporary Problems. Oceana Publications.

Shimm, Melvin G., ed., *Population Control: The Imminent World Crisis* (Dobbs Ferry, NY 1961).

Shubik, Martin (1970). A Curmudgeon's Guide to Microeconomics. *Journal of Economic Literature* 18, 2.

Soros, George (1997). The Capitalist Threat. *Atlantic Monthly*, February, 45-58.

Sullivan, Andrew (2007). Science is rescuing us from our moral mazes. *The Sunday Times* Review, 25 November, p. 11.

Symons, Michael (1987). In Adelaide, the collapse of the free market. Sydney, NSW: *Times on Sunday*, 3 May, p. 35.

Tarascio, Vincent J. (1986). The Crisis in Economic Theory: a Sociological Perspective. *Research in the History of Economic Thought and Methodology* 4, 283-95.

Tobin, J. (1987). On the Efficiency of the Financial System. *Lloyds Bank Review* 183 (July), 1-15.

Ukai, Yasuharu (1987). Cycles of Isolationism and Foreign Trade—a case study of Japan. *Kansai University Review of Economics and Business* 16, 1 (September), 61-74.

Wheeler, John A. (1965). Reference to Wynne-Edwards (1965), Private Communication 31 May, Palmer Physical Laboratory, Princeton.

Worldwatch Institute (2000). *State of the World 2000.* New York and London: W. W. Norton & Co.

Wynne-Edwards, V. C. (1965). Self-Regulating Systems in Populations of Animals. *Science* 26 March p. 1543.

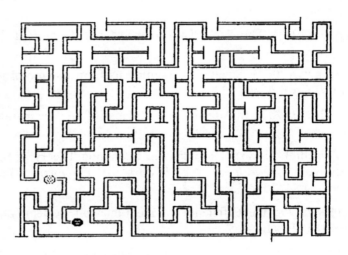

Index

page numbers followed by 'n' refer to footnotes

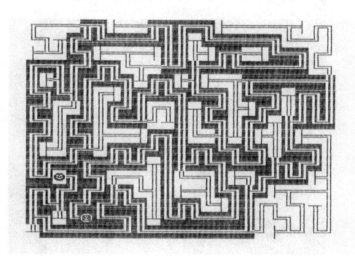

Sydney Afriat graduated in mathematics from Pembroke College, Cambridge, with a period during WW II in the High Speed Section, Aerodynamics Division, National Physical Laboratory, then DPhil at the Queen's College, Oxford. Work with Richard Stone in the Department of Applied Economics, Cambridge, 1953-56, initiated his activity in economics. After 1956-58 as Lecturer and Research Fellow in Mathematics, Jerusalem, years 1958-62 were in Princeton. Then Economics and Mathematics at Rice University, Houston, and Visiting Fellow at Yale. Beside periods at Purdue, UNC, Waterloo and Ottawa the later time includes intervals at Mathematical Sciences Research Institute, Berkeley, Visiting Fellow, All Souls College, Oxford, Izaak Walton Killam Memorial Fellow, Academic Visitor London School of Economics, Visiting Fellow, Macquarie University NSW, Visiting Professor, Institute of Social and Economic Research, Osaka, Professor, Bilkent University, Ankara, Visiting Professor, University of Siena, and Jean Monnet Fellow, European University Institute, San Domenico di Fiesole/Firenze; otherwise in UK and Italy; author of books and articles to do with mathematics and economics and scattered other items.

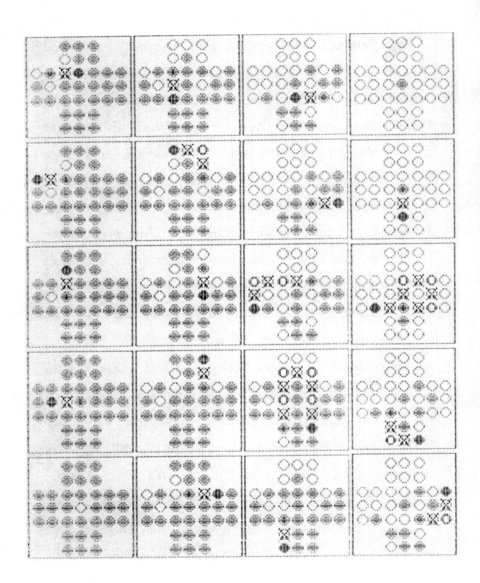